The Molecular World

Separation, Purification and Identification

edited by Lesley Smart

This publication forms part of an Open University course, S205 *The Molecular World*. Most of the texts which make up this course are shown opposite. Details of this and other Open University courses can be obtained from the Call Centre, PO Box 724, The Open University, Milton Keynes MK7 6ZS, United Kingdom: tel. +44 (0)1908 653231, e-mail ces-gen@open.ac.uk

Alternatively, you may visit the Open University website at http://www.open.ac.uk where you can learn more about the wide range of courses and packs offered at all levels by The Open University.

The Open University, Walton Hall, Milton Keynes, MK7 6AA

First published 2002

Edited, designed and typeset by The Open University.

Published by the Royal Society of Chemistry, Thomas Graham House, Science Park, Milton Road, Cambridge CB4 0WF, UK.

Printed in the United Kingdom by Bath Press Colourbooks, Glasgow.

ISBN 0 85404 685 2

A catalogue record for this book is available from the British Library.

1.1

s205book 8 i1.1

The Molecular World

This series provides a broad foundation in chemistry, introducing its fundamental ideas, principles and techniques, and also demonstrating the central role of chemistry in science and the importance of a molecular approach in biology and the Earth sciences. Each title is attractively presented and illustrated in full colour.

The Molecular World aims to develop an integrated approach, with major themes and concepts in organic, inorganic and physical chemistry, set in the context of chemistry as a whole. The examples given illustrate both the application of chemistry in the natural world and its importance in industry. Case studies, written by acknowledged experts in the field, are used to show how chemistry impinges on topics of social and scientific interest, such as polymers, batteries, catalysis, liquid crystals and forensic science. Interactive multimedia CD-ROMs are included throughout, covering a range of topics such as molecular structures, reaction sequences, spectra and molecular modelling. Electronic questions facilitating revision/consolidation are also used.

The series has been devised as the course material for the Open University Course S205 *The Molecular World*. Details of this and other Open University courses can be obtained from the Course Information and Advice Centre, PO Box 724, The Open University, Milton Keynes MK7 6ZS, UK; Tel +44 (0)1908 653231; e-mail: ces-gen@open.ac.uk. Alternatively, the website at www.open.ac.uk gives more information about the wide range of courses and packs offered at all levels by The Open University.

Further information about this series is available at www.rsc.org/molecularworld.

Orders and enquiries should be sent to:

Sales and Customer Care Department, Royal Society of Chemistry, Thomas Graham House, Science Park, Milton Road, Cambridge, CB4 0WF, UK

Tel: +44 (0)1223 432360; Fax: +44 (0)1223 426017; e-mail: sales@rsc.org

The titles in *The Molecular World* series are:

THE THIRD DIMENSION
edited by Lesley Smart and Michael Gagan

METALS AND CHEMICAL CHANGE
edited by David Johnson

CHEMICAL KINETICS AND MECHANISM
edited by Michael Mortimer and Peter Taylor

MOLECULAR MODELLING AND BONDING
edited by Elaine Moore

ALKENES AND AROMATICS
edited by Peter Taylor and Michael Gagan

SEPARATION, PURIFICATION AND IDENTIFICATION
edited by Lesley Smart

ELEMENTS OF THE p BLOCK
edited by Charles Harding, David Johnson and Rob Janes

MECHANISM AND SYNTHESIS
edited by Peter Taylor

The Molecular World Course Team

Course Team Chair

Lesley Smart

Open University Authors

Eleanor Crabb (Book 8)

Michael Gagan (Book 3 and Book 7)

Charles Harding (Book 9)

Rob Janes (Book 9)

David Johnson (Book 2, Book 4 and Book 9)

Elaine Moore (Book 6)

Michael Mortimer (Book 5)

Lesley Smart (Book 1, Book 3 and Book 8)

Peter Taylor (Book 5, Book 7 and Book 10)

Judy Thomas (*Study File*)

Ruth Williams (skills, assessment questions)

Other authors whose previous contributions to the earlier courses S246 and S247 have been invaluable in the preparation of this course: Tim Allott, Alan Bassindale, Stuart Bennett, Keith Bolton, John Coyle, John Emsley, Jim Iley, Ray Jones, Joan Mason, Peter Morrod, Jane Nelson, Malcolm Rose, Richard Taylor, Kiki Warr.

Course Manager

Mike Bullivant

Course Team Assistant

Debbie Gingell

Course Editors

Ian Nuttall

Bina Sharma

Dick Sharp

Peter Twomey

CD-ROM Production

Andrew Bertie

Greg Black

Matthew Brown

Philip Butcher

Chris Denham

Spencer Harben

Peter Mitton

David Palmer

BBC

Rosalind Bain

Stephen Haggard

Melanie Heath

Darren Wycherley

Tim Martin

Jessica Barrington

Course Reader

Cliff Ludman

Course Assessor

Professor Eddie Abel, University of Exeter

Audio and Audiovisual recording

Kirsten Hintner

Andrew Rix

Design

Steve Best

Carl Gibbard

Sarah Hack

Mike Levers

Sian Lewis

John Taylor

Howie Twiner

Library

Judy Thomas

Picture Researchers

Lydia Eaton

Deana Plummer

Technical Assistance

Brandon Cook

Pravin Patel

Consultant Authors

Ronald Dell (*Case Study:* Batteries and Fuel Cells)

Adrian Dobbs (Book 8 and Book 10)

Chris Falshaw (Book 10)

Andrew Galwey (*Case Study:* Acid Rain)

Guy Grant (*Case Study:* Molecular Modelling)

Alan Heaton (*Case Study:* Industrial Organic Chemistry, *Case Study:* Industrial Inorganic Chemistry)

Bob Hill (*Case Study:* Polymers and Gels)

Roger Hill (Book 10)

Anya Hunt (*Case Study:* Forensic Science)

Corrie Imrie (*Case Study:* Liquid Crystals)

Clive McKee (Book 5)

Bob Murray (*Study File*, Book 11)

Andrew Platt (*Case Study:* Forensic Science)

Ray Wallace (*Study File*, Book 11)

Craig Williams (*Case Study:* Zeolites)

CONTENTS

PART 1 CHEMISTRY: A PRACTICAL SUBJECT

Adrian Dobbs and Lesley Smart

PART 2 SPECTROSCOPY

Lesley Smart and Eleanor Crabb

CASE STUDY: FORENSIC SCIENCE

Andy Platt, Anya Hunt and Lesley Smart

Part 1

Chemistry: A Practical Subject

Adrian Dobbs and Lesley Smart

based on 'The Search for Purity'
by Keith Bolton and Malcolm Rose (1991)

INTRODUCTION: PREPARATION OF A COMPOUND

1

Chemistry is a fundamental science that underpins much of the world around us. It is also a practical subject. Although much of what we have learnt so far may have seemed conceptual or theoretical in nature, the basis for it has all come about through centuries of experimental laboratory work performed originally by individuals in their own homes, but nowadays by chemists — technicians, undergraduates, postgraduates and advanced researchers. None of the chemistry that you have learnt so far would have been known without these skilled experimentalists.

The aim of this book is to introduce you to many of the skills and techniques that are required by the modern chemist, such as how to perform a reaction, how to purify the products and finally how to prove your results — that you have actually made what you set out to make. In the text we can only describe the various procedures, but you will be able to watch many of them on the associated CD-ROM.

The skills and techniques described here are generally applicable to the whole of chemistry, whether it be an organic or inorganic experiment. Therefore rather than subdividing the book on the basis of the different branches of chemistry, we have integrated the material as far as possible, using examples from all areas of modern chemistry.

1.1 Planning a reaction

Before chemists can perform a reaction, just as in any profession, they need to *plan* exactly what they are going to do. If you were to ask practising chemists, they would all agree that time spent in planning a reaction is time well spent, and invaluable to the success of the experiment.

What are the major points which you should consider when planning a reaction? A list of most of the questions and points is given below.

- The scale of the reaction — how much product do you want to make?
- The mole ratios of the reactants; how much of each reactant to use?
- How expensive are the reagents? Are there cheaper alternatives?
- What is the most suitable solvent for the reaction?
- What temperature will be required?
- How long will the reaction take?
- Will you need to work under an inert and/or dry atmosphere?
- What equipment will be needed?
- Can the reaction be performed on the benchtop, or is a fume cupboard needed?
- What safety precautions will be necessary?

You also have to consider what you are trying to achieve during the reaction. Is the reaction probing some detailed reaction mechanism or is it preparatory — in other words, part of a long synthesis directed towards a desired product. An analytical chemist investigating a mechanism will have a very different set of priorities in planning a reaction compared to a synthetic chemist.

Chemists find that the careful keeping of a laboratory notebook is essential during their work. This involves carefully noting down everything that was done during an experiment from start to finish, recording relevant masses and other data such as temperature and timings, and noting *all* observations. If this is done in an orderly fashion, then it is very easy to draw conclusions from an experiment, to draw out data for a report or publication, to repeat the reaction, or simply to plan your next reaction.

An extract from a (rather idealized!) well-kept laboratory notebook should look something like Figure 1.1.

Notice the style and the various conventions that are used. The aim of the experiment and the equation for the reaction are set out clearly at the start, followed by the method and finally the results. A note is also made of any safety precautions necessary. Note that amounts of substances are placed in brackets after the compounds they refer to and are given in grams (or mls if the compound is a liquid) and also (preferably) in numbers of moles: this is conventional for formal reports and publications, so you may as well get used to it from the start.

Formal reports are always written in the past tense and the passive voice: '10 ml of water was added to the reaction' rather than 'I added 10 ml of water…'.

A template for how you should write-up your experiment in your laboratory notebook is given in Figure 1.2 (overleaf). You may well see variations on this style elsewhere and there is nothing wrong with most of these. However, if you follow this general format, you will not go far wrong when writing-up experiments.

1.2 Assembling the apparatus: doing the reaction

Before we can consider doing a reaction, we need to learn something about the apparatus that is available to use. You may have encountered some chemical apparatus before, for example a test tube, beaker or conical flask or even a bunsen burner. These alone however are insufficient to perform most reactions. Over the years, chemists have developed specialized apparatus for performing chemical reactions. In particular, we have glassware which is capable of withstanding extreme high and low temperatures and corrosive substances, and which can be used to keep out air and moisture. This specialized glassware consists of a series of interlocking tapered ground-glass joints (Figure 1.3 overleaf), which permit various pieces of glassware and apparatus to be connected together without the need for rubber stoppers, corks or any sort of rubber tubing connectors (the joints only need to be lightly greased). Collectively, this apparatus is known as **Quickfit® apparatus**, due to the easy and rapid way in which the apparatus may be connected and assembled.

Illustrated in Figure 1.4 (overleaf) is a typical set of glassware and Quickfit glass apparatus which you might encounter in any modern laboratory, whether it be in a university or in industry. You should try and familiarize yourself with the names and shapes of each of these pieces of apparatus, so that when you come to follow an experimental procedure, you know exactly what apparatus you need to assemble.

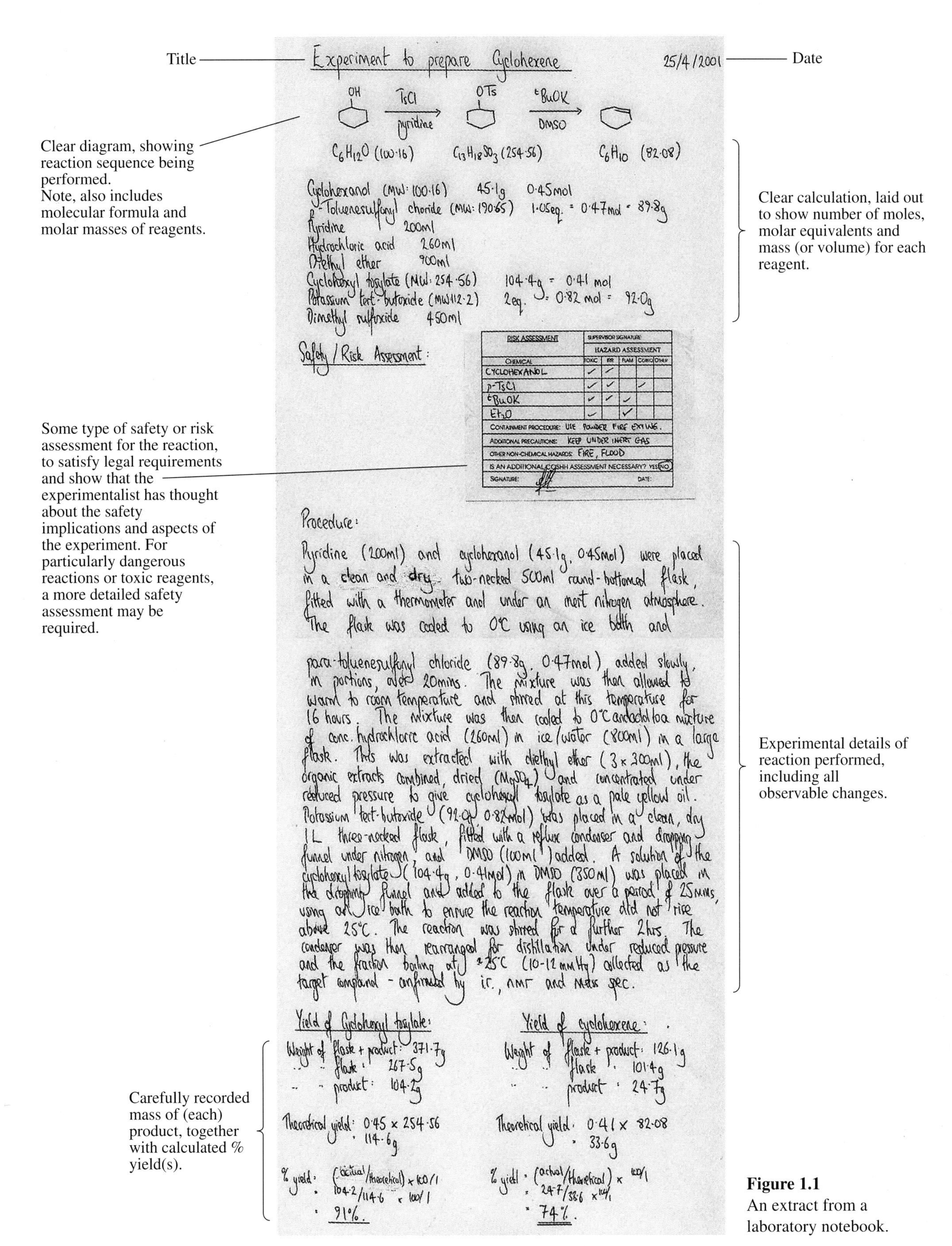

Experiment to prepare Cyclohexene — 25/4/2001

OH → (TsCl, pyridine) → OTs → (^tBuOK, DMSO) →

$C_6H_{12}O$ (100·16) — $C_{13}H_{18}SO_3$ (254·56) — C_6H_{10} (82·08)

Cyclohexanol (MW: 100·16) 45·1g 0·45mol
p-Toluenesulfonyl choride (MW: 190·65) 1·05eq. = 0·47mol = 89·8g
Pyridine 200ml
Hydrochloric acid 260ml
Diethyl ether 900ml
Cyclohexyl tosylate (MW: 254·56) 104·4g = 0·41 mol
Potassium tert-butoxide (MW 112·2) 2eq. = 0·82 mol = 92·0g
Dimethyl sulfoxide 450ml

Safety / Risk Assessment:

RISK ASSESSMENT	SUPERVISOR SIGNATURE:				
	HAZARD ASSESSMENT				
CHEMICAL	TOXIC	IRR	FLAM	CORRO	OTHER
CYCLOHEXANOL	✓	✓			
p-TsCl	✓	✓		✓	
^tBuOK	✓	✓	✓		
Et_2O	✓		✓		
CONTAINMENT PROCEDURE: USE POWDER FIRE EXTING.					
ADDITIONAL PRECAUTIONS: KEEP UNDER INERT GAS					
OTHER NON-CHEMICAL HAZARDS: FIRE, FLOOD					
IS AN ADDITIONAL COSHH ASSESSMENT NECESSARY? YES/NO					
SIGNATURE:				DATE:	

Procedure:

Pyridine (200ml) and cyclohexanol (45·1g, 0·45mol) were placed in a clean and dry two-necked 500ml round-bottomed flask, fitted with a thermometer and under an inert nitrogen atmosphere. The flask was cooled to 0°C using an ice bath and para-toluenesulfonyl chloride (89·8g, 0·47mol), added slowly, in portions, over 20mins. The mixture was then allowed to warm to room temperature and stirred at this temperature for 16 hours. The mixture was then cooled to 0°C and added to a mixture of conc. hydrochloric acid (260ml) in ice/water (800ml) in a large flask. This was extracted with diethyl ether (3 × 300ml), the organic extracts combined, dried ($MgSO_4$) and concentrated under reduced pressure to give cyclohexyl tosylate as a pale yellow oil. Potassium tert-butoxide (92·0g, 0·82mol) was placed in a clean, dry 1L three-necked flask, fitted with a reflux condenser and dropping funnel under nitrogen, and DMSO (100ml) added. A solution of the cyclohexyl tosylate (104·4g, 0·41mol) in DMSO (350ml) was placed in the dropping funnel and added to the flask over a period of 25mins, using an ice bath to ensure the reaction temperature did not rise above 25°C. The reaction was stirred for a further 2hrs. The condenser was then rearranged for distillation under reduced pressure and the fraction boiling at ≈25°C (10-12 mmHg) collected as the target compound – confirmed by i.r., nmr and Mass spec.

Yield of Cyclohexyl tosylate:
Weight of flask + product: 371·7g
" " flask: 267·5g
" " product: 104·2g
Theoretical yield: 0·45 × 254·56 = 114·6g
% yield: (actual/theoretical) × 100/1 = 104·2/114·6 × 100/1 = 91%.

Yield of cyclohexene:
Weight of flask + product: 126·1g
" " flask: 101·4g
" " product: 24·7g
Theoretical yield: 0·41 × 82·08 = 33·6g
% yield: (actual/theoretical) × 100/1 = 24·7/33·6 × 100/1 = 74%.

Figure 1.1
An extract from a laboratory notebook.

Title	**Experiment to prepare..........**	**Date.........**
Diagram of reaction including relative molecular masses/ formulae of reagents and products		
Name, relative molecular mass and amount used for each substance in reaction		
Safety/Risk Assessment	This is beyond the scope of the course, but you should be aware that any reaction requires a risk assessment to be performed prior to starting the experiment.	
Method (including all masses and amounts of materials used, both in g (or ml) and mols written in past tense		
All recorded data e.g. dry TLC plate, masses and yield(s), spectroscopic data	R_f = m.t. or b.t.: IR spectra: ^{1}H NMR: ^{13}C NMR:	

Figure 1.2 Template for an experimental write-up.

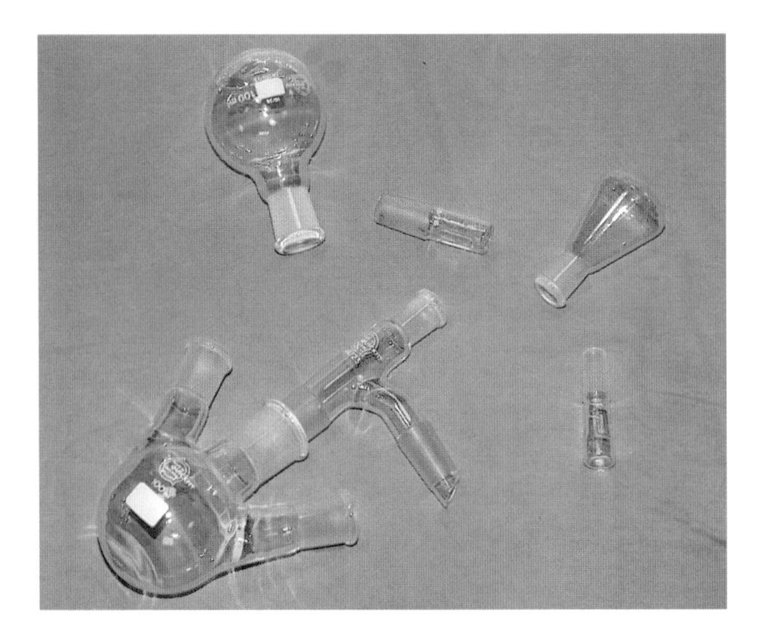

Figure 1.3
Quickfit glassware. Quickfit is a registered trademark of Bibby Sterilin Ltd.

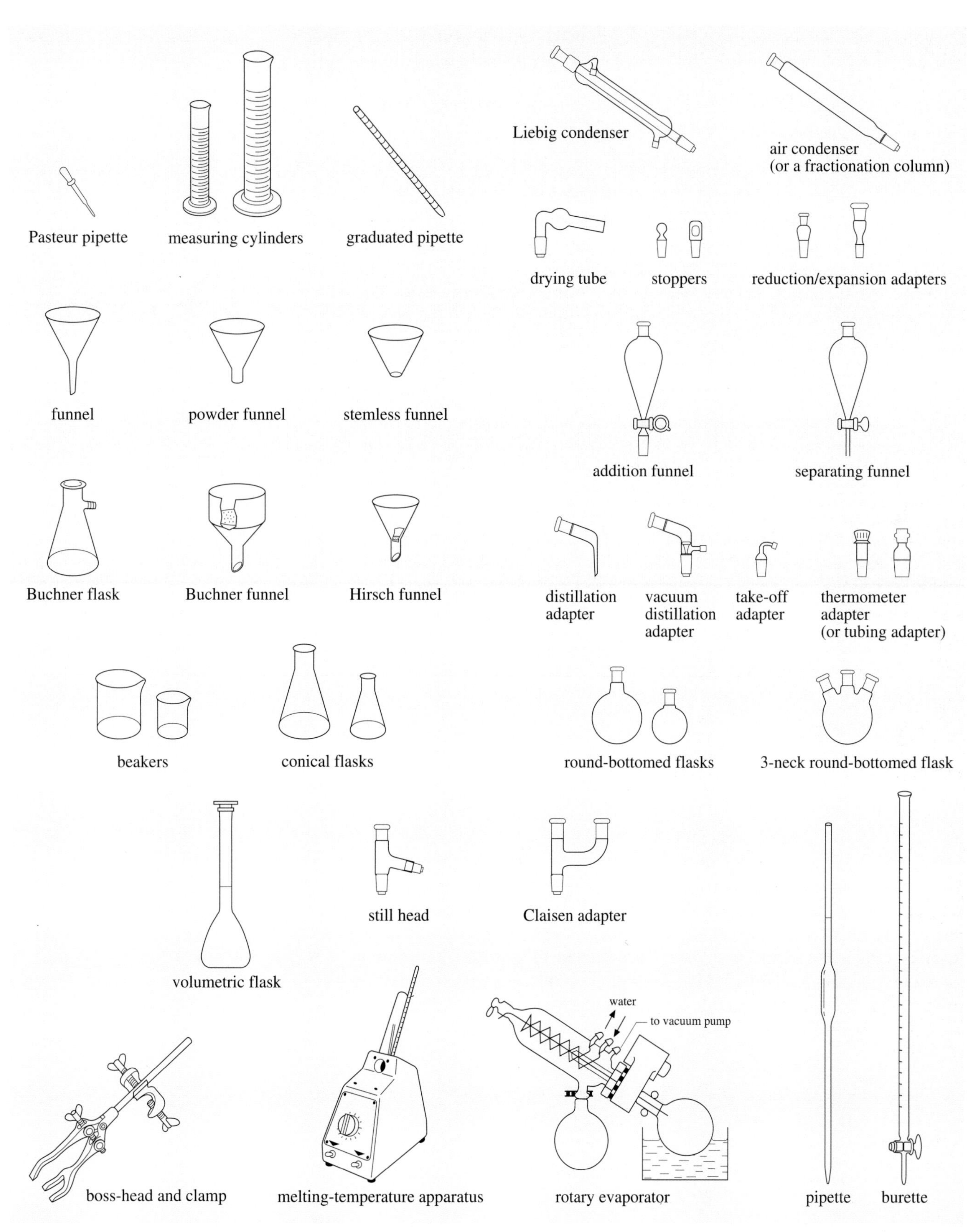

Figure 1.4 A selection of laboratory equipment.

Handling Quickfit apparatus is an acquired skill in its own right and it takes a while to be familiar with its use and capabilities. Quickfit apparatus comes in a variety of sizes, each perfectly adapted for large- or small-scale reactions. It is left to the experimentalist to decide which size of flask or funnel would be best for the particular reaction that is to be performed.

COMPUTER ACTIVITY 1.1 Looking at Glassware

At some point in the near future you should watch the video entitled *Looking at Glassware* in the multimedia activity called *Practical techniques* on the *Experimental techniques* CD-ROM that accompanies this book. This activity demonstrates how to assemble various pieces of chemical apparatus and illustrates the advantages of the interlocking Quickfit style of glassware. This activity should take approximately 10 minutes to complete.

Notice that when performing any experiment, normal laboratory safety procedures must *always* be followed, i.e. the wearing of protective clothing (usually a lab coat), safety spectacles, and gloves.

Under legislation known as **COSHH** (Control of Substances Hazardous to Health), a detailed **risk assessment** has to be made, documented and filed for every experiment performed. This may indicate that special safety precautions are deemed necessary, such as using a fume cupboard, or a face-mask. How to make these assessments is beyond the scope of this book *.

Once we have assembled the apparatus, we can start the reaction. Part of a typical experimental procedure may read as follows:

> 'Place 2-methylpropan-2-ol (25 g; 0.34 mol) and concentrated hydrochloric acid (85 ml†) in a 250 ml separating funnel and shake the mixture from time to time over 20 minutes.'

What exactly does this mean and how can we relate this to the apparatus you have just been learning about? From Figure 1.4, we can see what a separating (or separatory) funnel looks like. However, there are also certain assumptions in any given experimental procedure. For example, all apparatus should always be clamped securely (Figure 1.5a) so that it does not drop or fall over, and you may have noticed this as you watched *Looking at Glassware* in Computer Activity 1.1. This is always assumed rather than stated, as an experienced chemist knows that a separating funnel or round-bottomed flask cannot stand on its own. If we were to write this experimental procedure out in full, it is actually telling you to:

- Put on laboratory coat, goggles and gloves.
- Clamp a 250 ml separating funnel securely and close the tap.
- Place 2-methylpropan-2-ol (25 g; 0.34 mol) and concentrated hydrochloric acid (85 ml) inside the separating funnel, pouring them in carefully from a measuring cylinder, using a funnel.
- Place a stopper in the separating funnel and shake the mixture from time to time for a period of 20 minutes. Between each shaking, invert the funnel carefully,

* Risk assessments are considered further in *Exploring the Molecular World*[1].
† In practical work it is common to use 'ml' rather than the equivalent cm^{-3}.

holding the stopper tightly in place, and open the tap to release any excess pressure of gas. The reason for carrying out this last procedure, rather than the more obvious loosening of the stopper, is that if there is a pressure of gas inside the vessel, when you loosen the stopper it could blow hydrochloric acid fumes into your face. By inverting the funnel and releasing the gas through the tap, you can point it safely away from yourself. (Figure 1.5b).

- Based on your knowledge of the reactions of alcohols *, write an equation for the reaction being performed in the experiment we have just described.

- The experiment described the reaction of 2-methylpropan-2-ol with concentrated hydrochloric acid, the product that we hope to obtain is 2-chloro-2-methylpropane, via a nucleophilic substitution (S_N1) reaction mechanism.

$$(CH_3)_3C{-}OH \xrightarrow{H^+} (CH_3)_3C{-}\overset{+}{O}H_2 \xrightarrow{-H_2O} (CH_3)_3C^+ \xrightarrow{Cl^-} (CH_3)_3C{-}Cl \qquad (1.1)$$

This was a fairly simple experimental procedure. Another is described below, for the preparation of carbonatotetraamminecobalt(III) nitrate from cobalt(II) nitrate, ammonia, ammonium carbonate and hydrogen peroxide, by the unbalanced equation

$$Co(NO_3)_2 + NH_3 + (NH_4)_2CO_3 + H_2O_2 \longrightarrow [Co(NH_3)_4CO_3]NO_3 + NH_4NO_3 + H_2O$$

> 'Dissolve $(NH_4)_2CO_3$ (20 g; 0.21 mol) in distilled water (60 ml) and add concentrated aqueous ammonia (60 ml). While stirring, pour this solution into an aqueous solution of $Co(NO_3)_2$ (15 g; 0.052 mol, 30 ml of distilled water). Slowly add hydrogen peroxide (8 ml, 30% solution). Pour into an evaporating dish and concentrate to 90–100 ml over a bunsen burner (do not allow the solution to boil). During the evaporation time add $(NH_4)_2CO_3$ in small portions (5 g; 0.05 mol).'

This reaction would be done in a fume cupboard because ammonia fumes are extremely pungent and lachrymatory (they make you cry). No special equipment is required and it is a case of making a sensible choice of vessels for the mixing and heating. A fuller explanation of what we would actually do in each step of the procedure is:

- Weigh the solid $(NH_4)_2CO_3$ (20 g; 0.21 mol) using a top-loading balance and place in a 250 ml beaker.
- Measure 60 ml of water using a 100 ml measuring cylinder, add it to the beaker and stir with a glass rod to dissolve the solid.
- Measure 60 ml of conc. ammonia in the measuring cylinder and pour into the beaker carefully.
- Prepare the aqueous solution of $Co(NO_3)_2$ (15 g; 0.052 mol, in 30 ml of distilled water) similarly, in a small conical flask.
- Mount the conical flask on a magnetic stirrer, put in a magnet bar, and set the stirrer going.

(a)

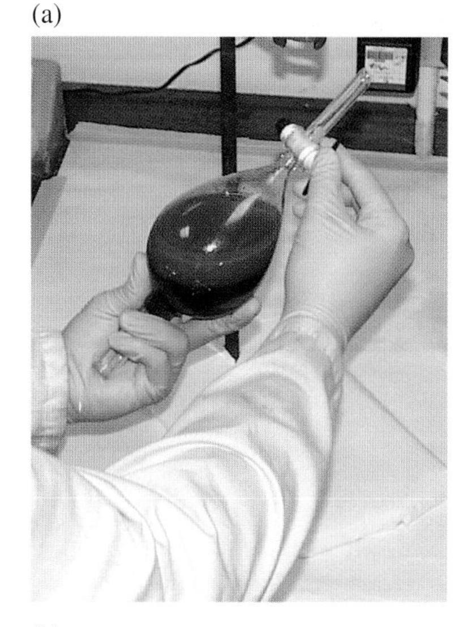

(b)

Figure 1.5
(a) Separating two layers using a separating funnel.
(b) Releasing pressure from a separating funnel.

* The reactions of alcohols is one of the subjects discussed in *Chemical Kinetics and Mechanism*[2].

- Add the ammonia solution from the beaker to the flask by pouring carefully either through a funnel or down a glass rod (Figure 1.6a).
- Measure the hydrogen peroxide (8 ml of a 30% solution) in a clean, dry 10 ml measuring cylinder and pour into the reaction mixture while maintaining stirring.
- Transfer the solution to a large evaporating dish which is supported over a bunsen burner using a tripod stand (Figure 1.6b).
- Heat very slowly and carefully to prevent spitting while gradually spooning in the previously weighed $(NH_4)_2CO_3$ (5 g; 0.05 mol) using a spatula.

Both these experimental procedures are comparatively straightforward, since no precautions have to be taken to exclude air, moisture, heat or light. Unfortunately, this is rarely the case and more often than not, chemists have to take specific precautions to exclude at least one of these factors, most commonly air (particularly oxygen) or moisture (as water vapour in the air). Our next experimental procedure shows the precautions that must be taken when performing a moisture-sensitive reaction — in this case the preparation of the organometallic complex $[\{Fe(CO)_2(\eta^5\text{-}C_5H_5)\}_2]$

$$2\,Fe(CO)_5 + C_{10}H_{12} \longrightarrow [\{Fe(CO)_2(\eta^5\text{-}C_5H_5)\}_2] + 6CO + H_2 \qquad (1.2)$$

C_5O_5Fe $\qquad$ $C_{14}H_{10}O_4Fe_2$

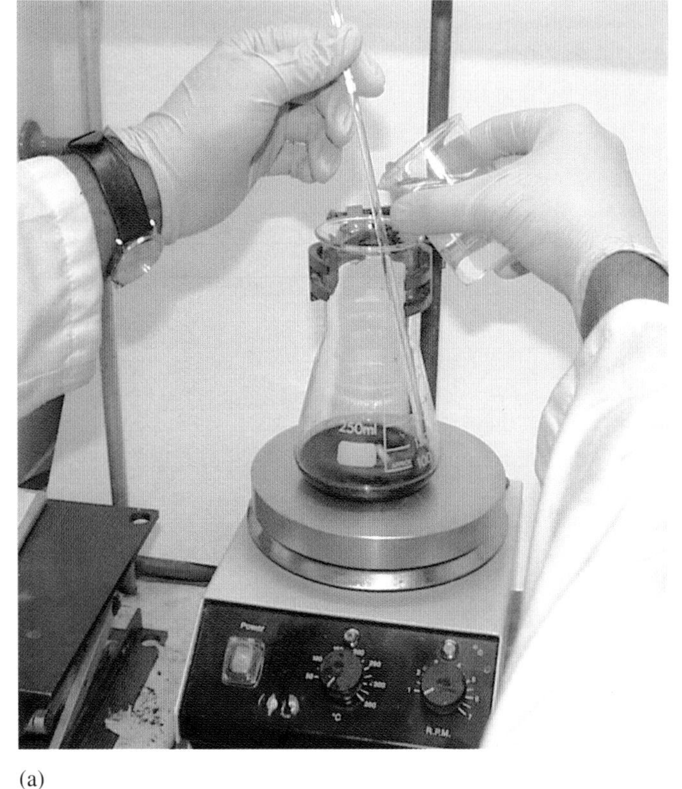

(a)

(b)

Figure 1.6
(a) Transferring a solution from one vessel to another.
(b) Evaporating a solution over a bunsen burner.

(Note that η^5 (pronounced 'eta five') refers to the way in which the C_5H_5 ring is bonded to Fe.)

> 'This procedure must be carried out in a fume cupboard. Assemble the apparatus shown in Figure 1.7, and perform the reaction under an atmosphere of dry nitrogen. Add $Fe(CO)_5$ (14.6 g, 10 ml; 70.5 mmol) and dicyclopentadiene (60 g, 64 ml; 455 mmol) to the flask. Reduce the nitrogen flow and heat the reaction mixture under reflux to 135 °C for 8 to 10 hours. (It is important not to let the temperature go below 130 °C (as no reaction will occur) or above 140 °C (decomposition of the product will occur). After the reaction period, allow the mixture to cool slowly to room temperature.'

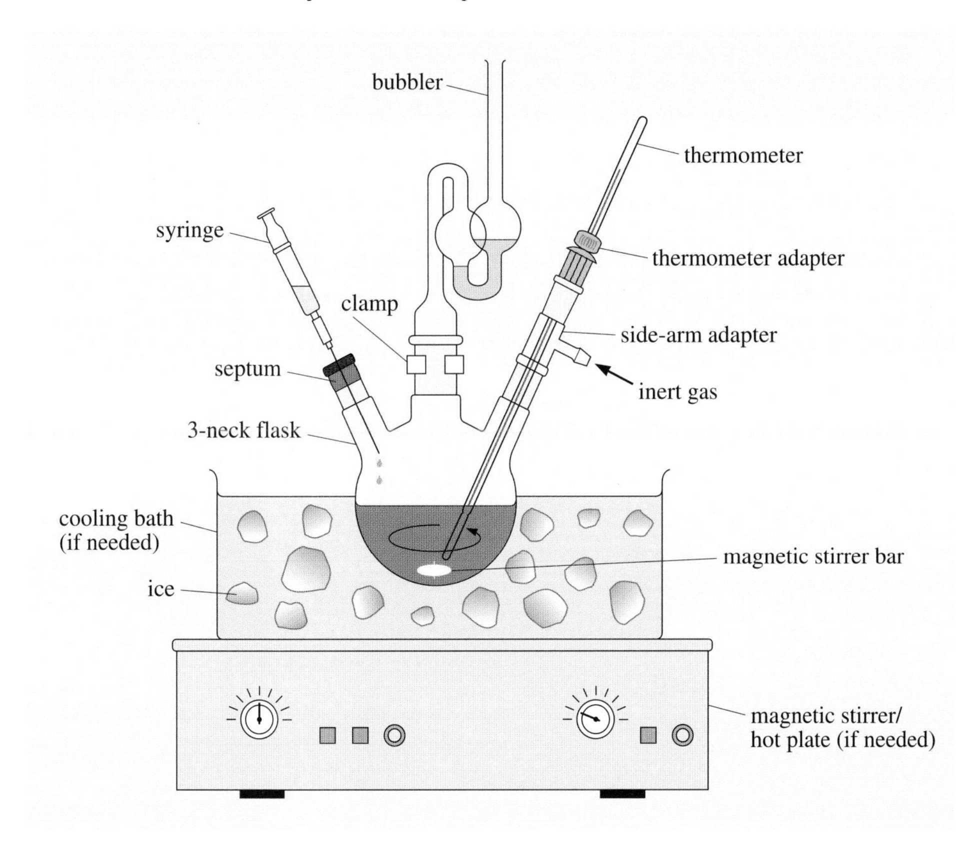

Figure 1.7
Typical arrangement of apparatus for an inert-atmosphere experiment. Note that the surrounding bath may be used to heat (oil bath) or cool (ice bath) a reaction. When heating, a reflux condenser would be placed in one of the necks of the reaction flask.

A fuller explanation of what we would actually do in each step in the procedure is:

- Assemble the apparatus as shown in Figure 1.7.
- Flush the system for 5 minutes with a rapid stream of nitrogen.
- With the nitrogen stream still flowing rapidly, remove the thermometer and add dicyclopentadiene (60 g, 64 ml, 455 mmol) to the round-bottomed 3-necked flask. To minimize your exposure to $Fe(CO)_5$, use a syringe to measure out and introduce the $Fe(CO)_5$ (14.6 g, 10 ml, 70.5 mmol) through the rubber septum into the flask. The constant stream of nitrogen will minimize air (which contains water vapour) entering the flask while the reactants are being added.
- Placing a reflux condenser between the flask and bubbler, turn the nitrogen flow down very low (one or two bubbles a minute). Using an oilbath, heat the reaction mixture under **reflux** to 135 °C for 8 to 10 hours. '*Heating under reflux*' means that you use a reflux condenser to prevent the volatile chemicals from escaping from the flask. The reflux condenser is cooled by circulating cold

water; when hot vapours rise up through it, they meet the cold surface, condense and drip back into the reaction mixture. The reaction temperature cannot rise above the boiling temperature of the solvent (you will see this demonstrated in Computer Activity 2.3).

- Carefully adjust the thermostatic control to maintain steady boiling, checking the temperature remains between 130 °C and 140 °C.
- After the reaction period, allow the mixture to cool slowly to room temperature, increasing the nitrogen flow slightly. (The nitrogen flow will prevent air from being drawn into the reaction vessel as it cools.)

We can immediately see that such a procedure is going to take a lot more time and will also require a great deal more care and skill from the experimentalist.

Practical chemistry is a manual skill in much the same way as cookery, woodwork or embroidery. It takes time and practice to learn and develop the right skills to be able to perform a reaction or synthesis with confidence.

Simply doing a reaction is not the end of the story; it is really just the beginning of a long process as we will see in the following sections.

1.3 Summary of Section 1

1. Chemistry is a practical subject, requiring specialist apparatus to perform most chemical reactions.
2. Chemists have a unique style of describing and writing-up experiments.
3. A risk assessment must always be made before performing any experiment.
4. It is sometimes necessary to perform reactions under a dry, inert gaseous atmosphere, to exclude all traces of moisture and oxygen.

QUESTION 1.1

The following is taken from a student's badly written laboratory notebook. Can you spot the mistakes and rewrite it in a proper scientific style?

'The three chemicals were put in a flask with a white plastic bar. A change had happened after 35 minutes, so I stopped the reaction and then added solvents and separated them. I evaporated one layer to give the product. The reaction was done under dry conditions. The starting material was a white solid and the product a yellow oil, which I got lots of.'

SEPARATING AND PURIFYING THE PRODUCT

2

Now that the reaction is over, you may think that the hard work is finished. Not so. Chemists often refer to the next stage of the process as the **work-up** of a reaction. A typical work-up procedure may be as simple as the addition of another reagent (in organic reactions this is often water or dilute acid) in order to finish or *quench* a reaction; but it may be a lengthy series of procedures, taking far longer than the actual reaction itself. Nevertheless, the work-up stage of a reaction may be critical to its success.

Sometimes the work-up of a reaction is simple. Take the case of the last of the three experiments that we considered, the preparation of the complex $[\{Fe(CO)_2(\eta^5\text{-}C_5H_5)\}_2]$ Here we saw a sophisticated experiment where we had to take precautions to exclude air and moisture in order to ensure success. This reaction has a very easy work-up procedure. After the reaction mixture has cooled to room temperature, deep-red crystals of the desired product form. The crystals are simply filtered off from the solution and dried in the air. This is an example of a complicated reaction where the desired product is the only solid obtained and so its isolation is very simple.

Now let us consider the reaction of a carbonyl group with a Grignard reagent (CH_3MgBr) to produce an alcohol *. You will often see this reaction written as

$$RC(O)R^1 \xrightarrow{CH_3MgBr} R(H_3C)C(OH)R^1 \qquad (2.1)$$

This is not quite a true representation of the experiment. The methyl group adds to the carbonyl group, but the product is not the alcohol, but a species known as a metal alkoxide (an ionic complex between the negatively charged oxygen atom and the positively charged metal)

$$RC(O)R^1 \xrightarrow{CH_3MgBr} R(H_3C)C(O^-\ {}^+MgBr)R^1 \qquad (2.2)$$

It is this alkoxide that is the product of Reaction 2.2 depicted above. The desired product, the alcohol, is only produced when dilute hydrochloric acid is added to the reaction flask at the end of the reaction, during *the work-up procedure*. So here, the work-up procedure is the second step of the reaction scheme.

$$RC(O)R^1 \xrightarrow{CH_3MgBr} R(H_3C)C(O^-\ {}^+MgBr)R^1 \xrightarrow{\text{dil. HCl}} R(H_3C)C(OH)R^1 \qquad (2.3)$$

Thus the correct way to write Reaction 2.1 would be

$$RC(O)R^1 \xrightarrow[\text{(ii) dil. HCl}]{\text{(i) } CH_3MgBr} R(H_3C)C(OH)R^1 \qquad (2.4)$$

* The use of a Grignard reagent is discussed in *Mechanism and Synthesis*[3].

As HCl is not the only acid that can be used, you will also see the abbreviations H^+/H_2O or H_3O^+ to mean 'add dilute acid'; all are perfectly acceptable alternatives.

The addition of an acid or a base to a reaction is a very common work-up procedure and it normally performs a dual function. It is the final step of a reaction sequence, liberating the desired product (such as forming the alcohol in the above sequence) while at the same time it destroys excess reagents and stops any further reaction taking place. This is particularly true of dry reactions, where we have taken special precautions to exclude all moisture from the reaction. Adding water to a dry reaction, destroys any excess moisture-sensitive reagent that may still be present, as well as performing any other work-up function (such as liberating the desired product in the above example).

- Why is it particularly important to exclude moisture from many reactions?
- Water (or water vapour in the air) can itself react with some of the reagents, thus destroying them.

In the above reaction, any moisture present would have destroyed the Grignard reagent.

Now we return to one of the other experiments that we considered in Section 1: the preparation of 2-chloro-2-methylpropane from 2-methylpropan-2-ol

$$(CH_3)_3C{-}OH + HCl \longrightarrow (CH_3)_3C{-}Cl + H_2O \qquad (2.5)$$

2-methylpropan-2-ol — 2-chloro-2-methylpropane

We saw that this is an easy experiment to perform. But how can we be sure that a reaction has taken place, and if the reaction has finished? Can we be sure that 2-chloro-2-methylpropane has been produced and not something else? Sometimes it is easy to know a reaction has occurred because something visible happens, for instance, the colour changes, or a precipitate forms, or a gas is evolved. In this case we only see a clear solution both before and after reaction. One possibility would be to test the reaction mixture for the presence of 2-chloro-2-methylpropane. But how? How do we identify the products of the reaction?

- What substances *could* be present in the reaction flask at the end of the preparation of 2-chloro-2-methylpropane?
- At the end of the preparation, we would hope that 2-chloro-2-methylpropane would be present. We would also expect to find some water (the other reaction product) and possibly some unreacted starting materials, HCl and $(CH_3)_3COH$. There may also be products of other possible side-reactions, such as elimination, or of reactions between the reactants and the products — in other words, we could have a complex mixture.

So, we do not just have to check that the desired product has been formed, but must also identify it from among a mixture of many possible components. As in most reactions, there is likely to be a variety of different substances present at the end of the reaction time, and the problem of identifying and isolating 2-chloro-2-methylpropane is not straightforward. Thus it is fortunate when crystals such as $[\{Fe(CO)_2(\eta^5\text{-}C_5H_5)\}_2]$ can be isolated so easily.

In order to complete a reaction, therefore, chemists require in their armoury:

- techniques for the separation of compounds from a mixture;
- methods for purifying separated components;
- a means of identifying which elements a compound contains;
- a method of determining the amount of each element in a compound (and thus determining the formula);
- a means of determining the structure of the compound.

There are very few techniques that allow a compound to be identified in the presence of many other compounds. Those that are available tend to be very specific and often expensive. So how do we proceed? We need to separate the components of the mixture *before* attempting an identification of each of them.

2.1 Solvent extraction and separation

In our experiment to produce 2-chloro-2-methylpropane, we reacted an organic compound (2-methylpropan-2-ol) with an aqueous mineral acid (HCl) and produced an organic product (2-chloro-2-methylpropane) and water. At the end of the reaction, the reaction flask probably contained a mixture of these four components, in varying amounts, along with some by-products. The first step towards purification that a chemist normally performs is a solvent extraction and separation.

The amount of a substance that dissolves in a particular solvent is the solubility of that substance.

> The **solubility** of a substance is defined as the number of grams of the substance that dissolves in 100 g of solvent under standard conditions.

Solubility is a physical property, and its value depends on the substance being dissolved, on the solvent and on the temperature. So at any particular temperature, different substances will have different solubilities in the same solvent. We can make use of this fact in the technique of **solvent extraction**.

BOX 2.1 Solubility

As most reactions are carried out in solution, it is useful to understand a little about solubility. Solubility is related to the molecular structure of both the solute and the solvent. If both are liquids then it is a matter of defining which you call the solvent and which the solute. A compound in a reaction which is not a reactant or a product or a catalyst is said to be the solvent. A liquid component in large excess, may be solvent as well as reactant.

The rule of thumb that chemists use for solubility is that 'like dissolves like'. This means that compounds with chemically similar structures tend to be miscible. For instance ethanoic acid (acetic acid), CH_3COOH, and methanoic acid (formic acid), HCOOH, are miscible in all proportions, that is they are completely soluble in each other.

Probably the most important factor in determining solubility is the polarity of solvent and solute, and this is closely related to the way that the electrons are distributed in chemical bonds, and throughout the molecule as a whole. If a bond has an uneven electron distribution then one atom in the bond bears a slight positive charge, and the other atom bears a slight negative charge; the bond is then said to be *polarized*. A molecule with an uneven overall electron distribution is called a **polar molecule**, and possesses an electric dipole moment such that it will align itself with an electric field. Water is a polar molecule. Alkanes and most compounds containing only carbon and hydrogen are almost non-polar. Generally organic molecules containing electronegative atoms such as oxygen and nitrogen are polar. Polar and non-polar solvents are often immiscible (imagine adding fat from a frying pan to water; the fat does not dissolve, but floats on the water).

Because like dissolves like, polar compounds dissolve in water, so carboxylic acids tend to be soluble in water, and ionic compounds dissolve in water or a polar solvent. Non-polar compounds such as alkenes are more soluble in non-polar solvents such as alkanes.

Problems start to arise with compounds such as octanoic acid which can be thought of as having a polar 'head' (the COOH group), and a 'non-polar tail' (the C_7H_{15} chain). Octanoic acid is insoluble in cold water and only slightly soluble in hot water, as the long alkyl tail dominates. Most organic molecules with a long hydrocarbon tail (more than 20 carbon atoms) are insoluble in water regardless of the functional group at the head. Table 2.1 lists common solvents in decreasing order of polarity.

Table 2.1 Common solvents, listed in decreasing order of polarity

Solvent	Formula	Dissolves ionic compounds	Degree of polarization
Water	H_2O	Yes	Most polar
Ethanoic acid (acetic acid)	CH_3COOH	Yes	
Methanol	CH_3OH	Yes	
Ethanol, etc.	ROH	Yes	
Propanone (acetone)	$(CH_3)_2C{=}O$	Yes	Intermediate
Ethyl ethanoate (ethyl acetate)	$CH_3COOC_2H_5$	No	
Ethoxyethane (diethyl ether)	$CH_3CH_2OCH_2CH_3$	No	
Trichloromethane, dichloromethane	$CHCl_3$, CH_2Cl_2	No	
Methyl benzene (toluene)	$C_6H_5CH_3$	No	
Benzene	C_6H_6	No	
Hydrocarbons	$CH_3(CH_2)_3CH_3$, etc.	No	Least polar

- In our synthesis of 2-chloro-2-methylpropane, assuming that there are only four components in the reaction flask (i.e. some unreacted starting materials and the two products), which do you think are likely to dissolve if we added (a) an organic solvent, (b) water?

- We have just seen that like dissolves like. The starting material, 2-methylpropan-2-ol and the product, 2-chloro-2-methylpropane will dissolve in an organic solvent, but not in water as they are non-polar molecules. If we add water, the unreacted HCl dissolves, but not the organic components.

Imagine now what would happen if we added equal volumes of an organic solvent and water. The two solvents will not mix, and so they will form two immiscible layers. The organic components will dissolve in the organic layer, and the ionic components will dissolve in the aqueous layer. By adding two different solvents to the reaction, one organic and one aqueous, we have a method for the separation of the organic and inorganic components of the reaction mixture. As we saw earlier, we do this by transferring the mixture into a separating funnel (Figure 1.5), adding water and an organic solvent, shaking well to mix the solvents thoroughly, leaving for a while to allow the layers to separate completely, and then running the bottom layer off through the tap into a flask or beaker, leaving the top layer behind in the funnel. This piece of apparatus makes separation of the two layers particularly easy. Typical organic solvents, which are immiscible with water and which are commonly used in solvent extractions are the ether, *ethoxyethane* (diethyl ether), *ethyl ethanoate* (ethyl acetate) and *toluene* (all of which have a lower density than water and thus float on water and form the top layer in a separating funnel) and *dichloromethane* (which has a greater density than water and thus sinks below water in a separating funnel). It is most important to remember which layer is which — many a time students have discarded the wrong layer and watched their precious compound disappear down the drain! Extraction and separation thus appears to be an ideal method for the separation of organic and inorganic materials.

Unfortunately, the extraction and separation are not always complete. The situation just described, where components are perfectly soluble in one type of solvent and completely insoluble in another, is the exception rather than the norm. Most substances are somewhat soluble in both organic and aqueous solvents, even if the solubility in one of these is particularly low. This feature is illustrated by the following example. Consider the preparation of pentane-1,5-dioic acid from pentane-1,5-dinitrile according to the following reaction

$$\underset{\text{pentane-1,5-dinitrile}}{NC(CH_2)_3CN} \xrightarrow[H_2O]{H_2SO_4} \underset{\text{pentane-1,5-dioic acid}}{HOOC(CH_2)_3COOH} \qquad (2.6)$$

If we were to perform an aqueous/organic extraction and separation at the end of this reaction, in which layer do you think that the product, pentane-1,5-dioic acid, will dissolve? The structure of this organic acid has both polar (two carboxylic acid groups) and non-polar (hydrocarbon chain) parts, neither of which dominates the other, and the acid dissolves in both polar and non-polar solvents, and so some of the product would be in the organic layer and some would be in the aqueous layer. Obviously, this is not very helpful to the chemist who would like a clean separation of the product into one layer or the other. Although the case of this acid is an extreme example, this is not an entirely unusual occurrence.

- How do you think the organic acid can be extracted from the solution in Reaction 2.6?
- Earlier, you were told that like dissolves like. Therefore, if we add an organic solvent to the mixture, the organic acid (together with any remaining starting dinitrile) should be extracted, to some extent, into this solvent, leaving the sulfuric acid, water and other polar substances in the aqueous layer.

- So if a proportion of the acid dissolves in the organic layer, how do you think we might be able to extract the remainder of the product from the aqueous solution?
- The simplest answer is to remove or separate the organic layer, and then add a further quantity of organic solvent and repeat the extraction.

In practice we do the extraction several times, as this maximizes the yield. After a reaction such as the synthesis of the dioic acid above, the aqueous solution is transferred to a separating funnel. Ethoxyethane (diethyl ether) is chosen as the organic solvent in this case. A 150 ml volume of ethoxyethane is added and, after shaking, the two liquids are allowed to separate. Ethoxyethane is almost insoluble in water, and a distinct boundary is observed between the water and the ether. The aqueous layer is run off into a separate container, and the remaining ethereal layer contains the organic acid. This organic layer (containing our product) is now stored in a separate container and the aqueous layer transferred back into the separating funnel. A further portion of ethoxyethane is added and the procedure repeated. Chemists normally go through this process three or four times as this ensures maximum efficiency in extracting the product from the aqueous layer.

- Why is it preferable to extract four times with small portions (e.g. 4×150 ml) of the ether, rather than extract once with a large volume (600 ml) of ether?
- Extracting with four portions of 150 ml of solvent is much more effective than extracting with a single volume of 600 ml. We know this from everyday experience; for instance, if we are extracting paint from paintbrushes, much cleaner brushes will result from washing with five portions of 20 ml of white spirit than with one portion of 100 ml of white spirit.

After extracting the aqueous solution from the reaction with four successive 150 ml portions of ethoxyethane, the ether extracts are combined. Water is not completely insoluble in ethoxyethane, and the small quantity of water present is removed from the ether by stirring it with a solid drying agent such as **anhydrous** magnesium sulfate. (Anhydrous means 'without water': not only is the magnesium sulfate dry, but it does not have any water of crystallization.) This is known as *drying the solvent*. The solid drying agent is removed from the ether solution by filtration. Finally, the solvent is removed from the dioic acid product by distillation over a hot water bath. (Distillation is covered in detail in Section 2.2.) The resulting residue in this case is fairly pure pentane-1,5-dioic acid.

QUESTION 2.1

A reaction mixture contains the following components:

Toluene; sodium chloride; benzyl bromide; ethoxyethane; potassium bromide and sodium hydrogen carbonate.

If equal volumes of water and ethoxyethane were added to this mixture in a separating funnel, what would be observed? Where would each component reside?

QUESTION 2.2

You should now be getting familiar with reading and understanding experimental procedures. Try and write an experimental and work-up procedure for the preparation of pentane-1,5-dioic acid from pentane-1,5-dinitrile.

The experimental procedure for the preparation of pentane-1,5-dioic acid is illustrated schematically in Figure 2.1.

Figure 2.1 The preparation of pentane-1,5-dioic acid.

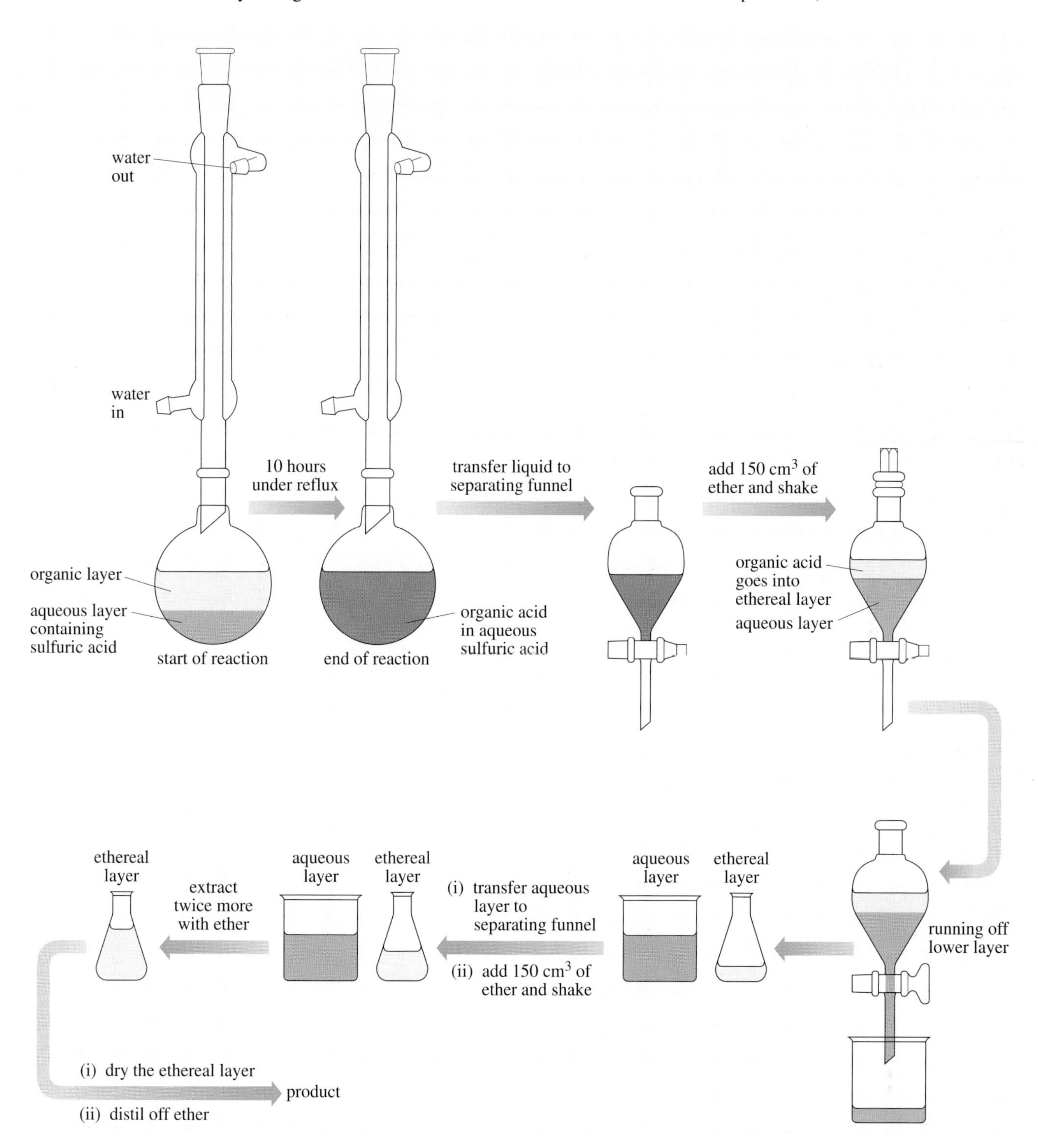

Extraction techniques were also employed in the preparation of another compound which we discussed earlier, 2-chloro-2-methylpropane. However, at the end of this reaction there were already two liquid layers present, without the addition of any extra solvent:

- an essentially organic layer, consisting of the unchanged alcohol and the product chloroalkane,
- an essentially inorganic, aqueous layer, containing hydrochloric acid, which becomes more dilute as the acid is used up and water is formed.

So here, a crude separation is straightforward, as shown in Figure 2.2; the two layers are allowed to settle and then the lower layer (the aqueous layer) is run off and set aside. The upper (organic) layer is then extracted again with water to remove traces of acid and other water-soluble impurities. (Notice that this is the reverse of the previous reaction: pentane-1,5-dioic acid was extracted from water using an organic solvent. Here we are using water to extract unwanted inorganic materials from an organic solvent.) Any residual water dissolved in the organic phase is removed by shaking it with anhydrous calcium chloride (a solid drying agent, like anhydrous magnesium sulfate). So we have rapidly achieved a separation of the organic components of the reaction mixture from the inorganic water-soluble ones. However, we are still faced with a common problem: the organic layer is still not the pure product (it contains at least two compounds — the starting material and the product), and further separation and purification are required.

The separation achieved using the extraction procedure is usually quite crude, but it often provides the first step in a series of purification steps, and can be particularly important where organic material needs to be separated from inorganic material. (Depending on the experiment, it may either be the organic or inorganic material that is required.) We shall return to the problem of the separation of the various components of our mixture shortly.

COMPUTER ACTIVITY 2.1
Solvent Extraction — one organic

At some point in the near future you should watch the video entitled *Solvent Extraction — one organic* in the multimedia activity *Practical techniques* on the *Experimental techniques* CD-ROM that accompanies this book. This activity deals with a case where the product to be extracted from the reaction mixture is that from a typical oxidation of a primary alcohol. Benzoic acid has been produced by the oxidation of benzyl alcohol with acidified potassium dichromate, and the organic product has to be separated from the remaining inorganic reagents. This activity should take approximately 10 minutes to complete.

$$\text{benzyl alcohol} \xrightarrow[H_2SO_4]{K_2Cr_2O_7} \text{benzoic acid}$$

Separation is also useful when an organic mixture contains a mixture of acidic, basic and/or neutral organic substances. A chemical reaction with an added basic or acidic solution can be used to effect a separation.

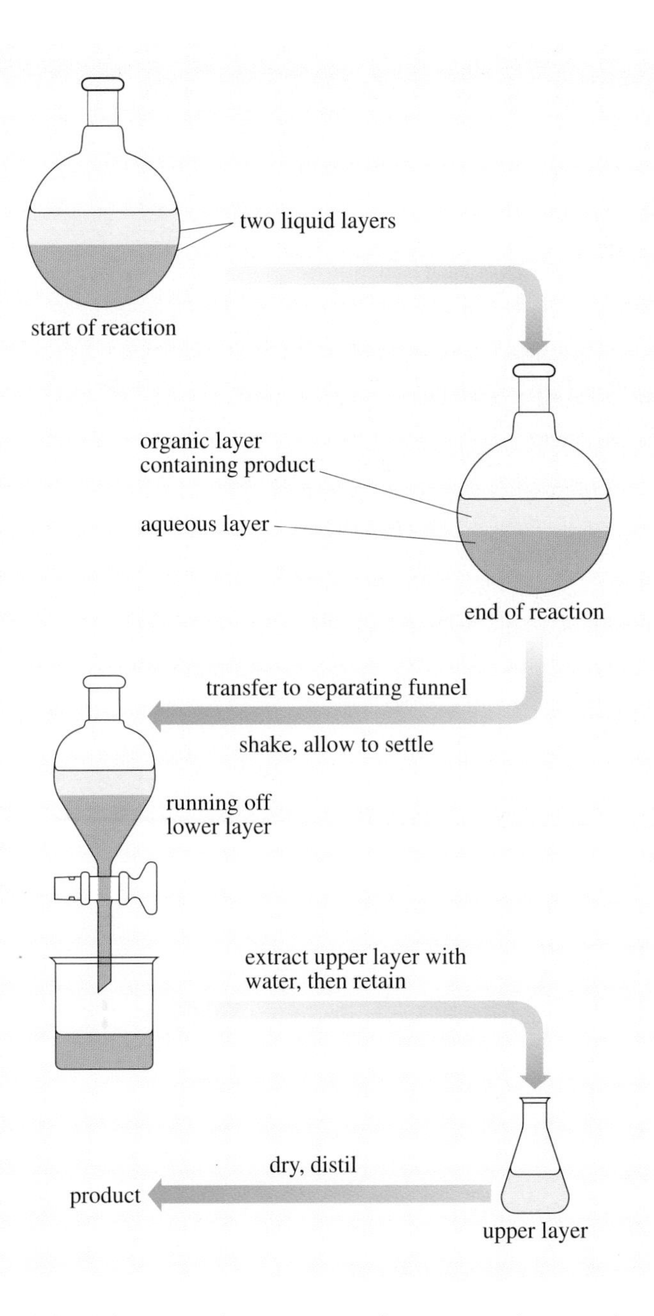

Figure 2.2
The preparation of 2-chloro-2-methylpropane.

In Computer Activity 2.2, we see how two organic compounds, an amine and a carboxylic acid, are separated from each other. This reaction mixture was formed by the hydrolysis of an amide

$$R^1C(=O)NHR^2 \xrightarrow{H^+/H_2O} R^1C(=O)OH \quad + \quad R^2\overset{+}{N}H_3$$

The first stage of the separation involves dichloromethane extraction of the acidic aqueous solution, containing the carboxylic acid and amine.

- Assuming that the aqueous acid was HCl, what chemical forms of the acid and amine will predominate in the 'acidified water'?

- The acid is in equilibrium with its *conjugate base*, R^1COO^-

$$R^1COO^- + H_3O^+ \rightleftharpoons R^1COOH + H_2O$$

The high concentration of H_3O^+ will force this equilibrium to the right, so the acid will exist predominantly in its neutral state.

The amine is in equilibrium with its *conjugate acid*

$$R^2NH_2 + H_3O^+ \rightleftharpoons R^2\overset{+}{N}H_3 + H_2O$$

The high concentration of HCl will force this equilibrium to the right also, so the amine will exist mostly in its conjugate acid form, $R^2\overset{+}{N}H_3$.

The alkyl ammonium ion, $R^2\overset{+}{N}H_3$, having a positive charge, will dissolve preferentially in the more polar of the two immiscible solvents, water. The less polar acid, R^1COOH, will prefer to dissolve in the solvent with lesser polarity, dichloromethane. So, by separating the solvents using a separating funnel, the carboxylic acid can be tapped off in the non-polar dichloromethane, leaving the amine behind in the acidic aqueous layer as its ammonium salt. After drying, evaporation of the dichloromethane then gives us the solid carboxylic acid.

The amine can be subsequently isolated by making the aqueous layer basic with aqueous sodium or potassium hydroxide, and then extracting the amine into ethoxyethane (diethyl ether).

- Give a chemical equation that describes this procedure.
- $R^2\overset{+}{N}H_3 + HO^- \rightleftharpoons R^2NH_2 + H_2O$

The equilibrium is forced to the right by a high concentration of hydroxide. Therefore, the amine will be predominantly in its neutral form under these alkaline conditions and, on the principle of like dissolves like, the amine will prefer a non-polar solvent like ethoxyethane. After separation and drying of the ethoxyethane layer from the aqueous layer, evaporation of the ethoxyethane gives the liquid amine. Now watch the procedures in the following Computer Activity.

COMPUTER ACTIVITY 2.2
Solvent Extraction — two organics

At some point in the near future you should watch the video entitled *Solvent Extraction — two organics* in the multimedia activity *Practical techniques* on the *Experimental techniques* CD-ROM that accompanies this book. This activity deals with the separation of an amine from a carboxylic acid. This activity should take approximately 10 minutes to complete.

The amine and acid could have been isolated in the reverse order. If the original mixture had been made alkaline first, the neutral amine would have been extracted by the non-polar solvent, and the acid would have remained behind in the water in the form of its highly polar conjugate base (R^1COO^-). Acidification of the water would then have returned the carboxylic acid to its neutral form (R^1COOH), allowing extraction into a non-polar organic solvent. This second approach is illustrated in Figure 2.3.

$$R^2\overset{+}{N}H_3 + HO^- \rightleftharpoons R^2NH_2 + H_2O$$

$$R^1COOH + OH^- \rightleftharpoons R^1COO^- + H_2O$$

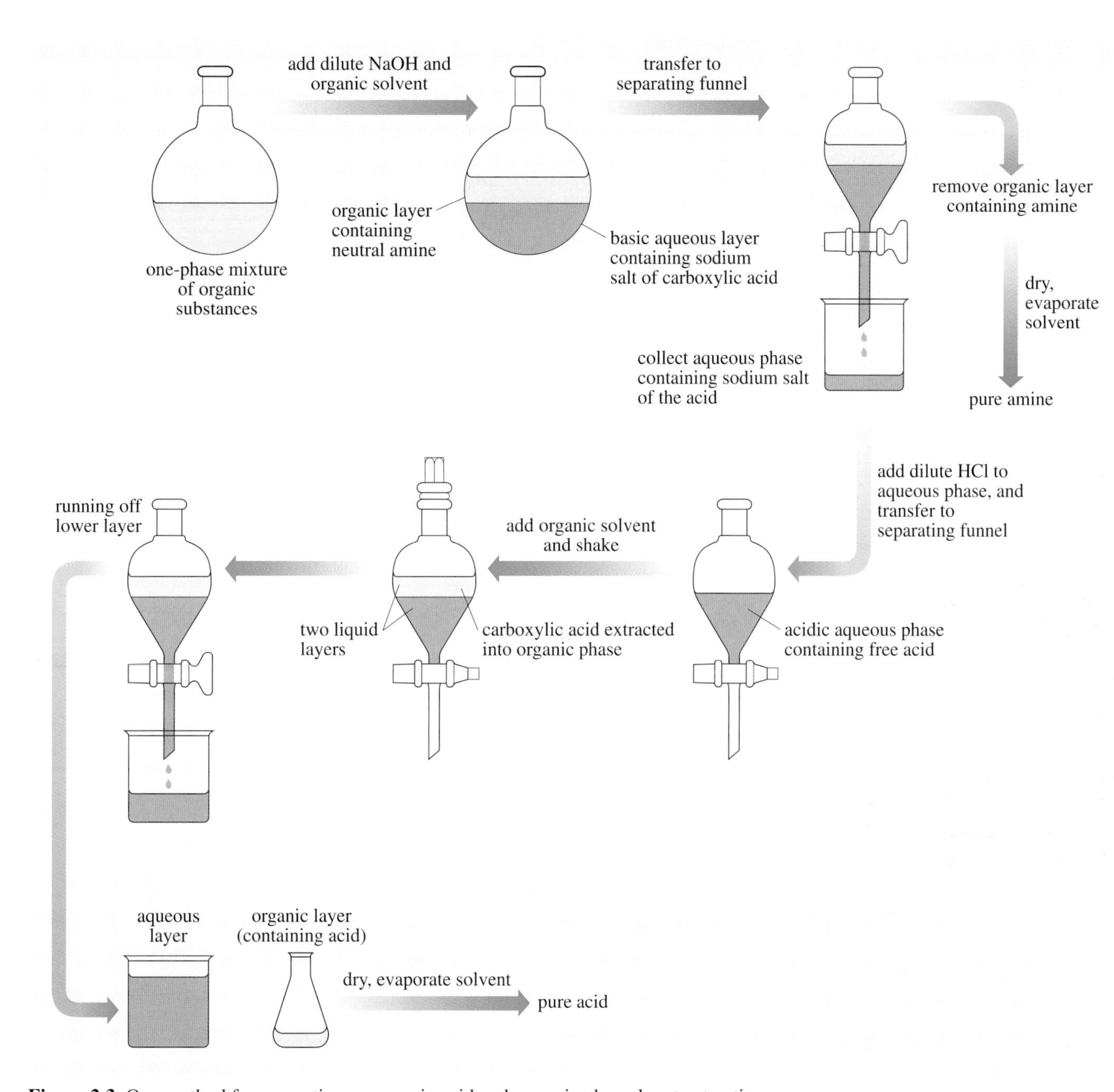

Figure 2.3 One method for separating an organic acid and an amine by solvent extraction.

Although the separations achieved using extraction procedures are crude, the technique can be applied to large volumes of mixtures, and is therefore important on both an industrial scale and on a laboratory preparative scale. In fact, in very simple mixtures, extraction procedures may provide the required degree of separation.

QUESTION 2.3

What is the rationale behind the separation schemes used in the following preparation?

1,5-dibromopentane can be prepared by the action of HBr in H_2SO_4 on pentane-1,5-diol

$$\mathrm{HO(CH_2)_5OH} \quad + \quad \mathrm{2HBr} \longrightarrow \mathrm{Br(CH_2)_5Br} \quad + \quad \mathrm{2H_2O}$$

pentane-1,5-diol 1,5-dibromopentane

Place a mixture of 47% hydrobromic acid (125 g; 85 ml) and concentrated sulfuric acid (37.5 g; 20.5 ml) in a 250 ml round-bottomed flask; add pure pentane-1,5-diol (17.5 g), attach a reflux condenser, and reflux gently for 2 hours. Allow to cool. Use a separating funnel to separate the lower layer of crude dibromide, and wash it with water (three 10 ml portions). Dry the product with anhydrous magnesium sulfate.

BOX 2.2 New 'Green' Solvents for Chemistry

By now you should be familiar with the idea of aqueous and organic liquids and the fact that they are immiscible — they do not mix. This is obvious from everyday examples in the world around us — think about drops of spilt petrol at the filling station floating on a puddle or the fat from a chip pan not mixing with water when you attempt to clean the pan. These are exactly the same phenomena as are occurring in a solvent separation. Unfortunately many of the organic solvents used traditionally for reactions and separations, such as benzene and carbon tetrachloride, are now known to be both toxic to humans and harmful to the environment. However, the use by the chemicals industry of such solvents still represents a huge market, with a current worldwide usage valued at £4 000 000 000 per annum. But their environmental impact is significant, and the Montreal Protocol has resulted in a compelling need to re-evaluate many chemical processes that were considered satisfactory for much of the twentieth century. Chemists throughout the world are searching for viable alternatives which will be less toxic and harmful to the environment. There are three leading candidates, all of which have their own particular advantages: *supercritical fluids*, *ionic liquids* and *fluorous solvents*.

Supercritical fluids are highly compressed gases that become liquid at temperatures at which they would normally be gaseous. They combine properties of gases and liquids in an intriguing manner. Supercritical carbon dioxide ($scCO_2$) has been the most extensively studied. Reactions are performed in a special high-pressure apparatus to maintain the gas in liquid form; when the reaction is complete, the pressure is reduced and the $scCO_2$ returns to the gaseous state and is then recycled. Nowadays coffee is decaffeinated using $scCO_2$ because this provides a safe means of removing the caffeine which dissolves in $scCO_2$, and leaves no harmful solvent residues in the coffee beans; any remaining CO_2 simply evaporates (Figure 2.4).

Ionic liquids are, quite simply, liquids that consist entirely of ions. Thus, molten sodium chloride is an ionic liquid (a solution of sodium chloride in water is an ionic solution). Many exist at room temperature, and are colourless liquids and easily handled. They are becoming increasingly important as media for synthetic organic chemistry (including polymer

Figure 2.4 Decaffeinated coffee.

chemistry and petrochemical processes), such that current commercial processes using conventional molecular organic solvents may be replaced by ionic liquids.

typical ionic liquids

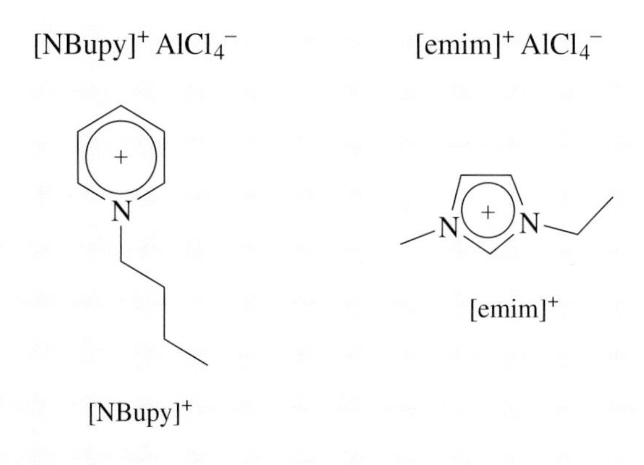

Finally, fluorous solvents are highly fluorinated, non-volatile, non-toxic liquids.

A fluorous solvent is based on a conventional solvent such as hexane, C_6H_{14}, where all the hydrogens have been replaced by fluorines (perfluorohexane, C_6F_{14}).

At certain temperatures, such fluorous solvents are immiscible with both organic solvents and water. The miscibility of the fluorous solvents with organic ones is highly-temperature dependent: they are immiscible at room temperature but miscible at higher temperatures (often only 50 to 60 °C). Analogues of conventional organic reagents are specifically designed to incorporate a '*fluorous tag*', i.e. a group such as C_6F_{13}, with a high fluorine content, that renders them preferentially soluble in fluorous solvents rather than organic ones. The reaction centre of the molecule however remains unaltered and unaffected. On performing reactions with fluorous reagents and solvents, organic substrates are introduced in an organic solvent and fluorous-tagged reagent(s) in a fluorous solvent. The two layers are immiscible and there is no reaction. However, on warming, the thermally dependent miscibility of the fluorous and organic solvents comes into play; the solvents are now miscible, they mix and the reaction proceeds as normal. The key comes on cooling again: the system returns to being biphasic, and the fluorous-tagged reagents and any fluorous-tagged by-products separate into the fluorous layer while the desired (untagged) products remain in the organic phase. A simple separation followed by evaporation of the organic solvent yields the desired product without any impurities being present.

Only time will tell if any of these alternative solvents fulfil their potential as a genuine replacement for traditional organic solvents.

a typical fluorous molecule

reaction centre P remains unaltered

three identical C_6F_{13} tags are introduced to each aromatic ring, thus rendering the molecule preferentially fluorous-soluble.

We must now move on to consider how we can separate the various components contained in one layer after solvent separation and washing. There are several possible techniques that we could use to separate mixtures of organic compounds, and all have the added advantage that they also purify the compound at the same time.

2.2 Separation by distillation

Distillation is a separation method that utilizes the different boiling points of the various components in a mixture to effect separation. Although distillation has been employed for centuries as a separation technique, the theory of the process for any but the simplest mixtures is extremely complex. However, here we are less interested in the theoretical aspects of distillation than in the factors that influence the technique as a tool for separation.

Let's consider the results of a simple distillation. At a pressure of 1 atm, pure benzene, C_6H_6, boils at 80 °C and pure methylbenzene (toluene), $C_6H_5CH_3$, boils at 110 °C. Now consider putting equal masses of benzene and toluene in the distillation apparatus shown in Figure 2.5 and slowly raising the temperature.

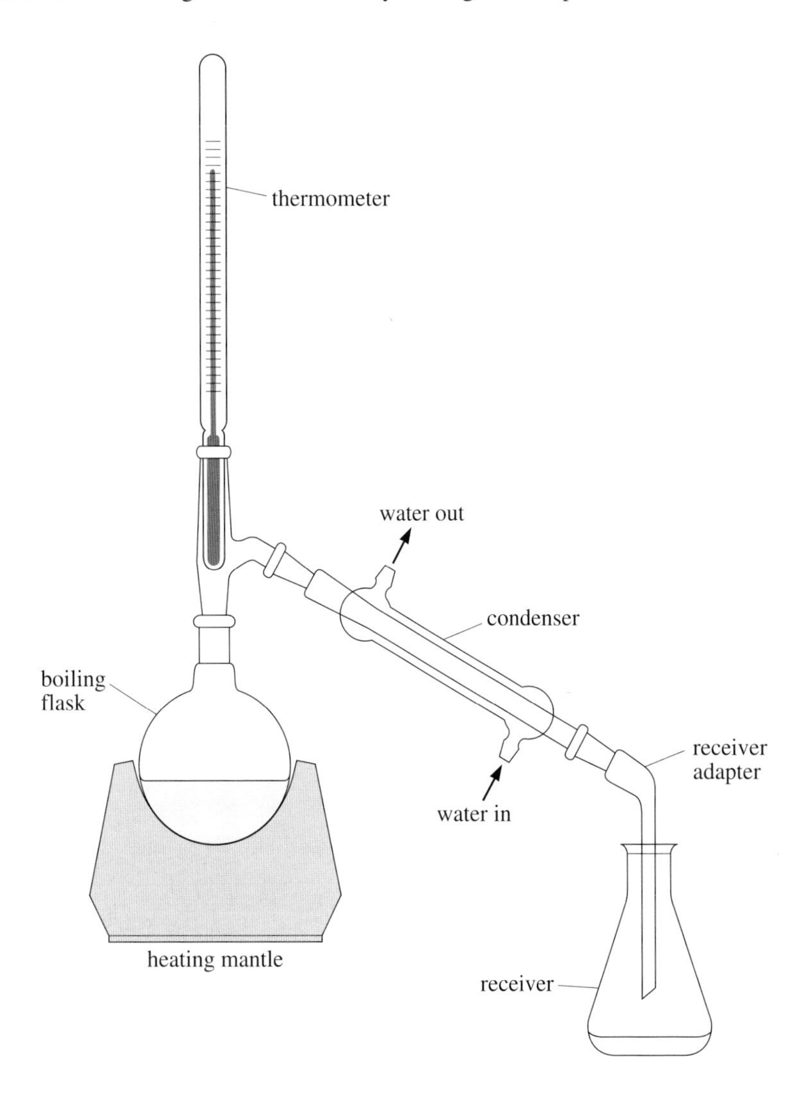

Figure 2.5
A simple distillation apparatus.

At what temperature do you think the mixture will start to boil?

(a) at 80 °C

(b) at 110 °C

(c) at a temperature midway between 80 °C and 110 °C

(d) at some other temperature

Answer (d) is correct. The mixture starts to boil at 92 °C; there is no simple relationship between the boiling temperature of the mixture and the boiling temperatures of the pure substances. (If you selected the wrong answer don't be too alarmed: there are still some modern chemistry textbooks that would tell you that the mixture would boil at 80 °C!)

However, as a separation technique, we are not so interested in the boiling temperature as in the nature of the liquid that is collected in the receiver.

In the above distillation, what do you think will be collected in the receiver?

(a) a liquid that has the same composition as the original

(b) a liquid that is richer in benzene than the original

(c) a liquid that is less rich in benzene than the original

(d) virtually pure benzene

The correct answer is (b).

You may be surprised at the separating efficiency; the first drop of liquid collected consists of 70% (by mass) of benzene and 30% (by mass) of toluene; that is, we have produced a 70 : 30 mixture from a 50 : 50 mixture. As you can see, the separation is not ideal. Furthermore, the first drop of liquid provides the best separation; as distillation proceeds, the boiling temperature increases as the relative amount of toluene distilling over increases. So, even for components with a difference in boiling temperature as great as 30 °C, the separation is poor. Such a procedure is useful only when there is a very large difference between the boiling temperatures of the components that are to be separated.

However, this separation can be improved. If we take the first sample of mixture that is collected in the receiver and repeat the distillation just on this small amount of liquid, then from this second run the first drop of liquid to distil over consists of 85% (by mass) of benzene. By repeating the process again, a further improvement in separation can be obtained. This time-consuming, repetitive process can be carried out in one piece of apparatus — a distillation apparatus known as a *fractionating column* (Figure 2.6 overleaf).

This **fractional distillation** greatly increases the speed and efficiency of the separation process. The column is packed with glass beads or some other inert material which has a large surface area. The vapour from the boiling liquid can condense on the surface of the inert material and can then be boiled again by the hot vapours coming up the column as the condensing liquid runs down. In this way the distillation process is repeated many times *within* the column. The column is therefore equivalent to many single distillation systems such as that shown in Figure 2.5. If the packing material and the length of the column are carefully chosen, fractionating columns are capable of efficiently separating liquids with boiling temperatures only 2 °C apart.

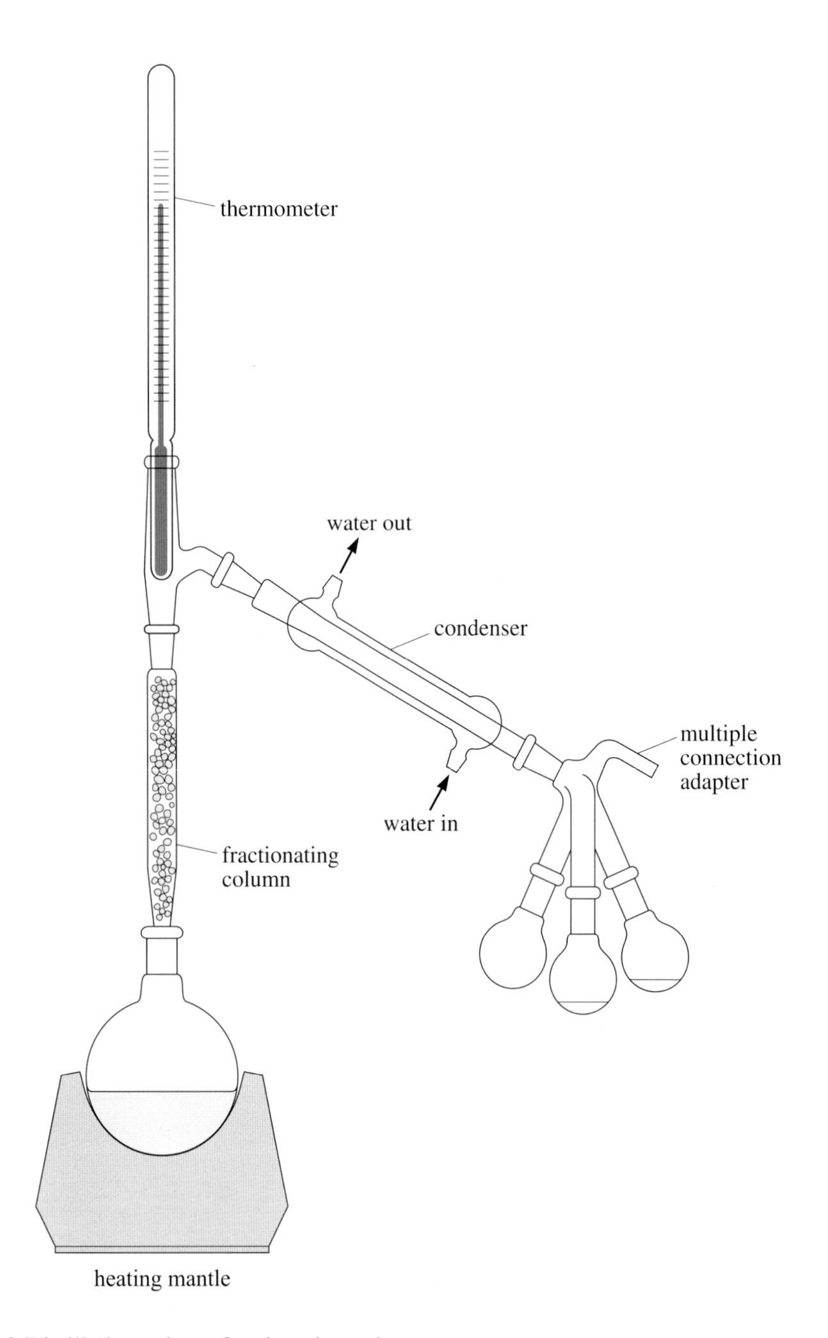

Figure 2.6 Distillation using a fractionating column.

BOX 2.3 An example of distillation on a large and practical scale

Much of the drinks industry relies on distillation as a key stage in the preparation of alcoholic beverages. Distillation is crucial to the preparation of whisky — hence the reason that these plants are often known as distilleries (Figure 2.7a). The key ingredients of whisky are water, peat and barley, and the sources of these and the types of soil that they come from are said to be the key in producing the variations in flavour between the differing brands of malt whisky.

Fresh barley is soaked in water for two days, then drained and spread out in a warm chamber to initiate germination — the *malting* process. As the barley starts to shoot, the insoluble starch in the seed is converted into a soluble sugar. This takes around seven days.

The germination process is halted by *smoking* the barley. The peat is burnt to give a thick, dark and rich smoke, which is then passed through the barley for eighteen hours. The smoked barley is dried in a smoke-free oven for twelve hours and then allowed to 'rest' for twenty eight days in storage.

The seeds are then mixed with pure water at 67 °C, which releases the sugars formed during the germination process. The sweet water, called *wort*, is cooled and allowed to ferment via the addition of yeast. Once the alcohol (ethanol) level has reached 8.5%, the *wash* is transferred for distillation.

All distilleries have differing shapes of distillation flasks (Figure 2.7b), and all will claim that theirs is the best! The whisky is normally distilled twice: once in the wash-still to produce the *low*, taking the concentration of alcohol up to 22%, and then a second time as the low contains impurities. This takes the alcohol level to 68%. It is the *middle cut* or middle section of this second distillation which becomes the whisky beloved by many.

This process is shown on the first part of the video in Computer Activity 2.3 which follows.

(a)

(b)

Figure 2.7 (a) Whisky distillery, (b) whisky stills.

COMPUTER ACTIVITY 2.3 Distillation

At some point in the near future you should watch the video entitled *Distillation* in the multimedia activity *Practical techniques* on the *Experimental techniques* CD-ROM that accompanies this book. This activity should take approximately 10 minutes to complete.

Distillation is very good for separating large volumes of liquid mixtures containing a small number of components, but it is not useful for separating each component of a complex organic mixture. Distillation is used industrially to separate the different fractions of crude petroleum in the oil-refining process.

Figure 2.8
Industrial fractionating columns used in an oil refinery.

- The statements in the preceding paragraph might appear to be contradictory, but in fact they are not. Why?
- Distillation of crude oil on an industrial scale requires a vast array of giant fractionating columns (Figure 2.8), but even then the separation is only partial.

For example, the gasoline fraction of crude oil contains all the many compounds with boiling temperatures between approximately 40 °C and 150 °C, but further separation is not required for most uses. The internal combustion engine runs better on a mixture of compounds than it does on any individual component. So distillation is useful for separating the components of the complex mixture of crude oil into approximate boiling ranges, but the separation is far from complete.

If we now return to our synthesis of 2-chloro-2-methylpropane, it turns out that a fractional distillation is the method of choice for the final separation of the product of the reaction. If you recall, we had successfully separated (by washing techniques) the organic and inorganic materials. We were left with two organic compounds, 2-methylpropan-2-ol (starting material) and 2-chloro-2-methylpropane (product), in an organic solvent. By performing a fractional distillation and collecting the boiling fraction between 49 °C and 51 °C we are able to obtain a pure sample of the desired product. Therefore the entire experimental procedure should read

> 'Place 2-methylpropan-2-ol (25 g; 0.34 mol) and concentrated hydrochloric acid (85 ml) in a 250 ml separating funnel and shake the mixture from time to time during 20 minutes. Draw off and set aside the lower acid layer. Wash the organic layer with water (20 ml). Dry the organic layer with anhydrous calcium chloride (5 g). Distil, collecting the fraction boiling between 49 °C and 51 °C.'

Notice again how in written experimental procedures, the fine details such as the assembly of glassware, and filtering off the solid drying agent, are all left out from the written instructions and are simply assumed knowledge; however, the experimentalist must still perform these steps.

We have now completed our synthesis of 2-chloro-2-methylpropane. We have performed a reaction, carried out a work-up and separated our product. In this case the final separation technique, fractional distillation, has also produced a pure product, so no further purification step is needed. All that remains for the chemist now is to confirm that the pure product obtained is actually the product that we set out to make originally. How this is done is the subject of Section 5. But before we do this, we must consider some other separation and purification techniques, for cases where we cannot use distillation (e.g. if the product is a solid).

Finally, what would happen if the boiling temperature of our desired fraction is very high (e.g. 160 °C)? It might well be dangerous to heat the mixture of compounds to such a high temperature. Even if your target compound is stable at these high temperatures, some of the other components in the mixture may not be, and may even decompose dangerously or explosively. Therefore, chemists have developed a technique known as distillation under reduced pressure. *By lowering the pressure, any liquid will boil at a lower temperature.* This is a relatively straightforward procedure; by reducing the pressure in the distillation apparatus using a pump, the boiling temperatures of *all* the components in the mixture are considerably lowered. Providing there is still a reasonable difference in the reduced boiling temperatures, separation will still be effected.

It turns out that distillation as a separation technique, is very much the domain of the organic chemist rather than the inorganic. One important exception is the cross-over area of organometallic chemistry: for example, many tin-containing compounds, such as tetraethyltin and tri-*n*-butyltin hydride, are high boiling-temperature liquids, which are separated from the reaction mixture and purified by distillation. Tetraethyltin is prepared by the reaction of tin tetrachloride with ethylmagnesium bromide

$$SnCl_4 + 4C_2H_5MgBr \longrightarrow Sn(C_2H_5)_4 + 4MgBrCl$$

and the product is the fraction boiling at 180 °C to 182 °C under normal atmospheric pressure.

It is interesting that as soon as some of the organic part of the molecule is replaced, as in diethyltin dichloride, the compound is a solid

$$SnCl_4 + Sn(C_2H_5)_4 \longrightarrow 2Sn(C_2H_5)_2Cl_2$$

QUESTION 2.4

What is the rationale behind the separation scheme used in the following preparation?

Pentanoic acid is prepared by the hydrolysis of pentanenitrile according to the following reaction

$$CH_3(CH_2)_3CN + 2H_2O \xrightarrow{\text{NaOH}} CH_3(CH_2)_3COOH + NH_3$$

Place pentanenitrile (15.0 g; 15.8 ml) and a solution of pure sodium hydroxide (14 g) in water (40 ml) in a 250 ml round-bottomed flask, and reflux until the pentanenitrile layer disappears (5 to 10 hours). Allow to cool, add water (15 ml) then slowly, and with external cooling, add 50% (by volume) sulfuric acid (20 ml). Separate the upper layer of pentanoic acid and dry it with anhydrous calcium sulfate. Distil and collect the pentanoic acid at 183 °C to 185 °C.

BOX 2.4 Can you boil an egg up a mountain?

As altitude increases, the air becomes 'thinner' — its density decreases — and the pressure drops. Pressure from the atmosphere helps compress water molecules into a liquid, but as soon as the pressure drops, it becomes easier for the water molecules to vaporize, and the boiling temperature of water decreases with decreasing pressure (Figure 2.9).

The answer to the question 'can you boil an egg up a mountain?' has therefore to be yes; but the boiling temperature of the water will drop as you go higher. An egg cooks when the protein in the albumen starts to denature, which we see when it changes from clear to white. This only begins to happen above about 60 °C to 65 °C. Because of the reduced boiling temperature of water as the pressure drops, it becomes very difficult to *cook* an egg at high altitudes!

If it takes 5 minutes to hard-boil an egg at sea-level, we can expect that it will take considerably longer on the top of Ben Nevis!

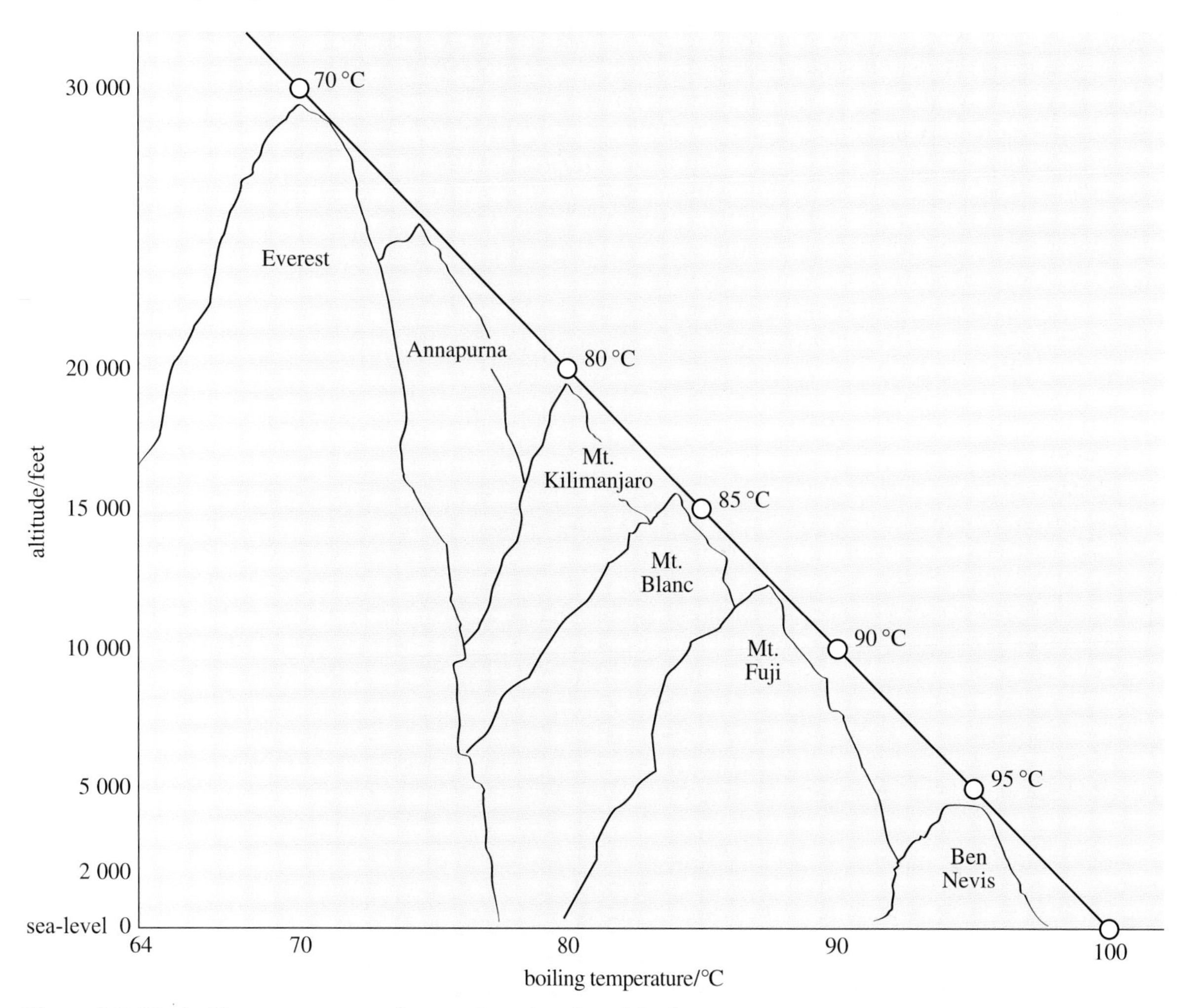

Figure 2.9 The boiling temperature of water plotted against altitude.

2.3 Chromatography

We now introduce a new technique known as **chromatography**, which can separate the components of quite complex mixtures. Chromatography is probably the most useful method for separating compounds to purify and identify them. There are many different forms of chromatography, and we shall start by concentrating on two types, **thin-layer chromatography**, usually called **TLC**, and **column chromatography**. In both of these methods, we use a stationary, solid phase over which flows a mobile, liquid phase. The separation works on the principle that each of the components in a mixture will have a different polarity and will adsorb (**adsorption**, refers to the ability of a substance to adhere to the surface of a solid) onto the stationary phase, the **adsorbent**, and dissolve in the mobile phase to a different extent. Thus, each of the components of the mixture will be pulled along by the mobile solvent at different rates.

The most commonly used adsorbents for both TLC and column chromatography are silica gel and alumina. Silica gel (SiO_2) is a general purpose adsorbent useful for a broad range of organic and ionic compounds. Alumina (ultrafine aluminium oxide, Al_2O_3) is available in acidic, basic and neutral forms. The basic form is used to separate basic and neutral compounds that are stable to base. The basic form is the most active, the neutral less active (but very good for separating strongly adsorbing groups like ketones and esters) and the acidic form the least active of all, but very useful for separating acids.

2.3.1 Thin-layer chromatography

In thin-layer chromatography (TLC), the solid phase, the adsorbent, consists of many small particles attached to a flat plate (which can be glass, plastic or metal foil) in a very thin layer; this is known as the *TLC plate*. A small amount of the reaction mixture to be separated is dissolved in the minimum amount of a solvent that dissolves all components of the mixture. A small spot of the mixture is then applied to the plate about 10 mm from the bottom (Figure 2.10a). (This process can be repeated several times to ensure there is a sufficient amount of material for separation.) The plate is then placed in a covered glass container containing a different solvent which is called the **eluant** (Figure 2.10b). The solvent slowly rises up the silica gel (a process known as **elution**) and, if a suitable solvent has been chosen, the compounds move up the plate at different rates, and the mixture will begin to separate on the plate as they gradually move up the plate behind the solvent. When the solvent has moved about three-quarters of the way up the plate, the plate is removed from the solvent and the position of the solvent front marked quickly before the plate dries (Figure 2.10c). The distance from the starting line to the solvent front can be measured, as can the distance from the starting line to the centre of each spot. A **retardation factor**, the R_f value, can then be calculated for each spot using the following equation

$$R_f = \frac{\text{distance of spot from origin}}{\text{distance of solvent front from origin}}$$

Using the values given on the left of Figure 2.10c, calculate the R_f values for the pink and blue spots.

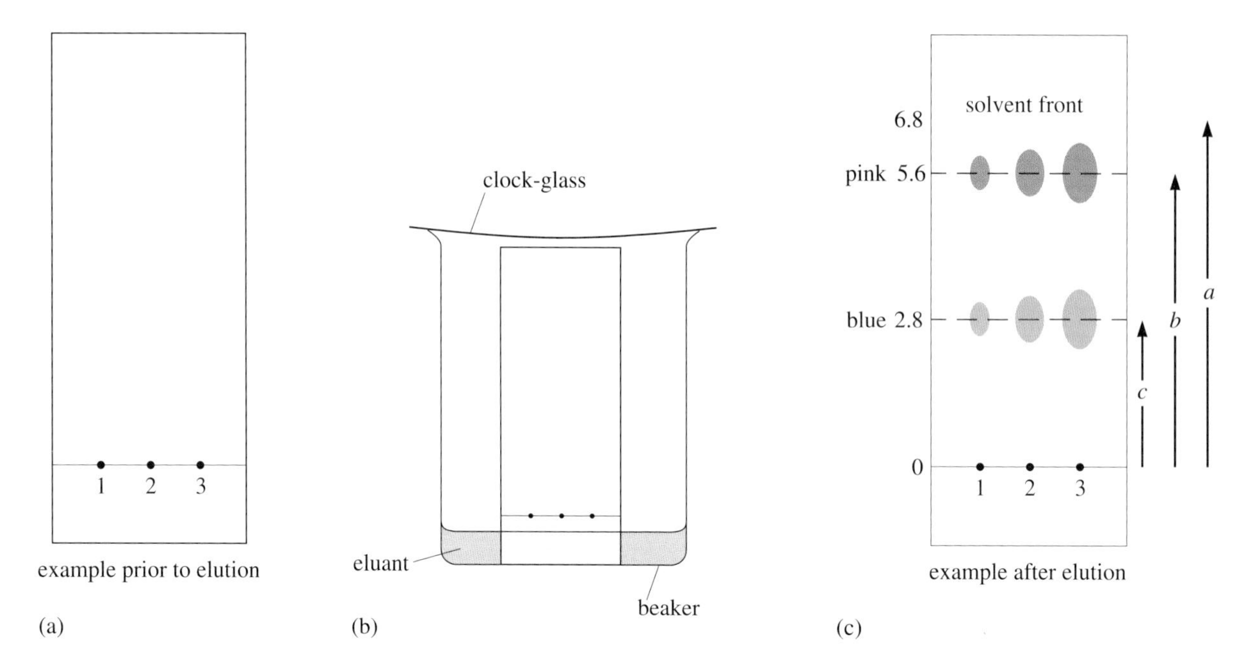

Figure 2.10 (a) A TLC plate before elution (spot 1 contains one application, spot 2, two applications and spot 3, three applications); (b) Running a TLC chromatogram; (c) Example after elution.

$$R_{\rm f}(\text{pink}) = \frac{b}{a} = \frac{5.6}{6.8} = 0.82$$

$$R_{\rm f}(\text{blue}) = \frac{c}{a} = \frac{2.8}{6.8} = 0.41$$

COMPUTER ACTIVITY 2.4 Thin-Layer Chromatography

At some point in the near future you should watch the video entitled *Thin-Layer Chromatography* in the multimedia activity *Practical techniques* on the *Experimental techniques* CD-ROM that accompanies this book. This activity should take approximately 5 minutes to complete.

A difficult choice is always that of which solvent to use to 'run' or elute the TLC plate. There is no easy answer to this question. Chemists choose using a combination of experience and trial-and-error; they would normally run several TLC plates, each using a different solvent or mixture of solvents, and find which gives the best separation. Table 2.2 lists common solvents in order of increasing polarity, as a guide to solvent selection.

Say we have a strongly polar reaction mixture on the plate. If we elute with petroleum ether or hexane, we have a non-polar solvent running over the very polar silica gel — these of course are opposites and so the solvent is not held strongly to the adsorbent. When the solvent encounters the reaction mixture, which *is* strongly attracted to the adsorbent, the non-polar solvent will not be able to displace it and so the mixture will not move. If instead, we elute with methanol which is highly polar, the solvent also adsorbs strongly to the adsorbent and so will displace almost every molecule that it encounters and everything in the reaction mixture moves together at

Table 2.2 Common solvents listed in order of increasing polarity

Solvent	Polarity
Petroleum ether and hexane	Least polar
Cyclohexane	
Methylbenzene (toluene)	
Dichloromethane, trichloromethane (chloroform)	
Ethoxyethane (diethyl ether)	Increasing polarity
Ethyl ethanoate (ethyl acetate)	
Propanone (acetone)	
Ethanol	
Methanol	
Water	Most polar

the solvent front. In this case neither of the two extreme solvents, hexane or methanol would effect a good separation. The interesting solvents are those with intermediate polarity. For most organic separations, chemists usually start with hexane and then gradually introduce a more polar solvent, such as ethoxyethane or ethyl ethanoate, in varying amounts until good separation is achieved.

TLC is an ideal technique for monitoring the progress of a reaction, where you do not want to lose large quantities of material from the reaction mixture as this would diminish the overall reaction yield. At the start of a new preparative reaction, a chemist would typically run a TLC of the starting materials. At regular intervals, they would sample the reaction mixture and again run a TLC of this mixture alongside samples of all the starting materials. As the reaction proceeds and starting materials are converted to products, the spots in the reaction mixture due to starting materials should gradually disappear, while new spots should appear elsewhere on the plate — hopefully corresponding to the desired product. When all the starting material spots have disappeared, the chemist knows that the reaction is complete and can stop the reaction and work it up. Therefore by using TLC we have answered one of the fundamental questions that we posed at the start of this book — how do we know when a reaction is complete?

The main drawback with TLC, is that it can only be performed on a very small scale, and so is not useful for separating the entire reaction mixture into its various components — for this we need a large-scale version of TLC, which we discuss in the next section. Now try doing a TLC experiment for yourself in the next Computer Activity.

COMPUTER ACTIVITY 2.5 Thin-layer chromatography in use: an application from the food industry

At some point in the near future you should watch the video entitled *Thin-layer chromatography in use: an application from the food industry* in the multimedia activity *Practical techniques* on the *Experimental techniques* CD-ROM that accompanies this book. There you will see an experiment on the separation of food colourings. At various times you will be asked to take notes or make measurements from the screen, so you should make sure that you have an experiment notebook and pen to hand. This activity should take about

STUDY NOTE

Open University students taking SXR205 *Exploring the Molecular World* should now study Section 2 of that Course, which covers this experiment in more detail.

20 minutes to complete.

The structural formulae and European list numbers of the food colours used are shown in Figure 2.11.

When you have finished, try to answer Questions 2.5 to 2.7 below.

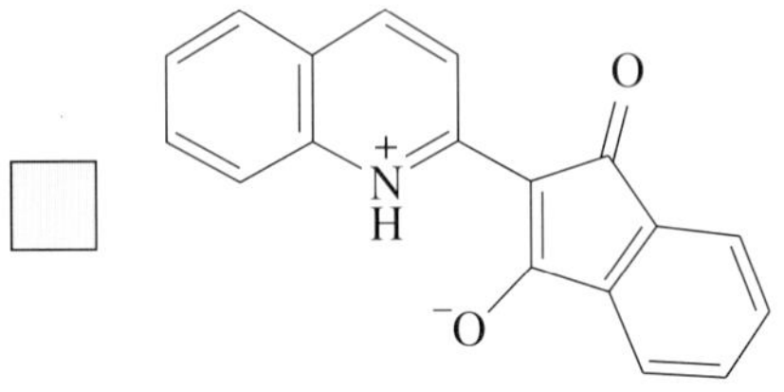

Quinoline Yellow (E104), bright yellow

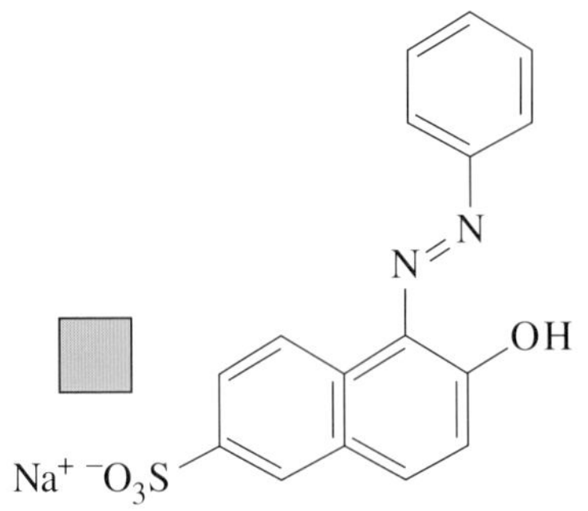

Sunset Yellow (E110), orange yellow

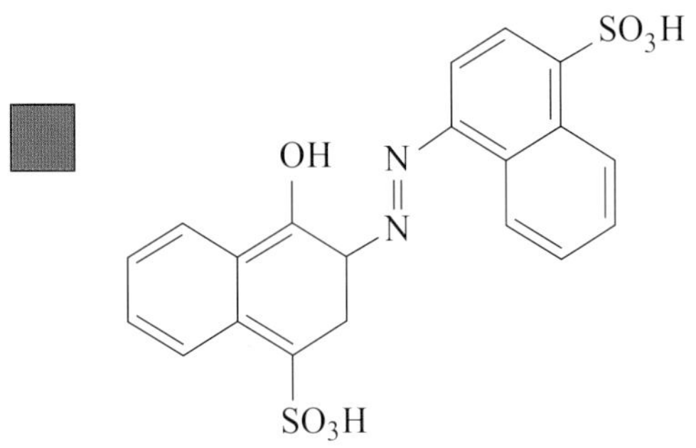

Carmoisine (E122), dark red

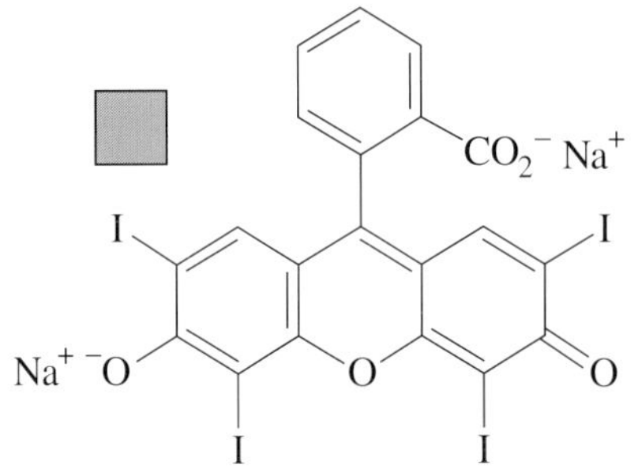

Erythrosine (E127), pink-red

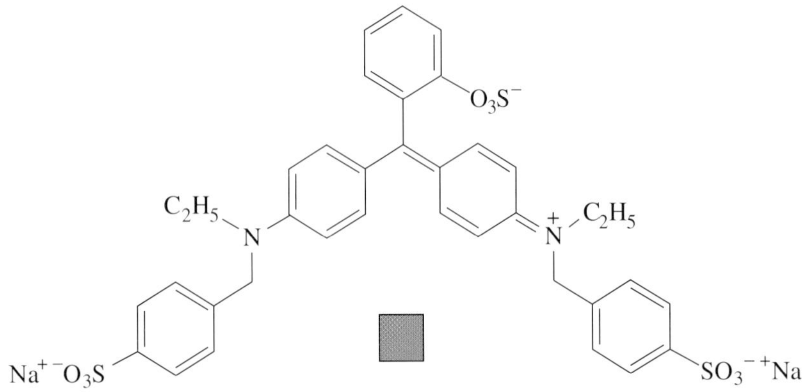

Brilliant Blue (E133), bright blue

Figure 2.11 The structural formulae and European list numbers of the food colours in Computer Activity 2.5.

The compounds we used in the experiment in Computer Activity 2.5 were coloured, and the spots were easily visible on the plate. To make a colourless compound visible on a TLC plate, we would have to allow the compound to interact, while it is being adsorbed on the surface of the plate, with something that will bring about a colour change. Many organic compounds are oxidized by potassium permanganate, which is itself reduced to manganese dioxide (brown). Silica gel does not react with potassium permanganate, so oxidized compounds show up as brown spots on a purple background when the plate is sprayed with dilute potassium permanganate solution. Many organic compounds adsorb iodine vapour, giving brown spots, or show up as bright (or dark) spots under the light from an ultraviolet lamp.

QUESTION 2.5

R_f values of between 0.2 and 0.8 are considered satisfactory. Why do you think values outside this range are not as good?

QUESTION 2.6

Is thin-layer chromatography very sensitive to the amount of mixture applied in the spots? Is this an advantage or a disadvantage?

QUESTION 2.7

How do you think the technique of thin-layer chromatography could be useful to (i) a chemist working in an analytical laboratory; (ii) a chemist trying to isolate and identify compounds from natural sources (a natural products chemist); (iii) a chemist preparing complex compounds from simpler starting materials (a synthetic chemist)?

2.3.2 Column chromatography

Thin-layer chromatography can only be used to separate small amounts of a mixture. If we want to separate larger amounts, we clearly need to use more of the solid phase and the solvent. This is achieved using column chromatography. Here, the solid particles are packed into a column, and the solvent flows down through the particles by gravity. The mixture is put on top of the column and, as the solvent flows through the column the different components move down (with the solvent) at different rates. Each component flows out of the other end of the column at a different time. By collecting the solvent in portions called **fractions**, we can isolate each component of the mixture as it comes out of the column. By changing the solvent running through the column, we can increase the polarity of the mobile phase, and thus remove the more polar components in turn from the column.

COMPUTER ACTIVITY 2.6 Column Chromatography

At some point in the near future you should watch the video entitled *Column Chromatography* in the multimedia activity *Practical techniques* on the *Experimental techniques* CD-ROM that accompanies this book. There you will see a video of this technique using aluminium oxide (alumina) as the stationary phase. This activity should take approximately 5 minutes to complete.

Let's now look at the chromatographic process in a little more detail, as illustrated schematically in Figure 2.12 (overleaf). Different substances are adsorbed to different extents on a particular material. The plates used for TLC in Computer Activity 2.4 were coated with a very fine layer of silica, a polar material. As a mixture of substances, dissolved in a suitable solvent, passes over the silica, different substances become adsorbed onto the silica surface to different extents. Polar organic compounds will be more strongly adsorbed than non-polar organic compounds, and so the progress of polar organic compounds up the silica surface will be slower. Note that the separation is taking place at the molecular level; each time a small collection of molecules of the mixture comes into contact with the surface of the grains of silica, adsorption is possible. This is why the chromatographic technique is so powerful at separating the components of a mixture, compared with distillation and solvent extraction. We saw in Section 2.2.1, how much more efficient the extraction technique is if, instead of using one batch of solvent, we use the same amount of solvent to extract the organic material in several smaller batches. As the number of extractions increases, so the separation improves. In chromatography the separation is conducted at the molecular level, so we are essentially increasing to the maximum the number of opportunities for separation. You could say that chromatography is equivalent to solvent extraction using a vast number of portions of solvent.

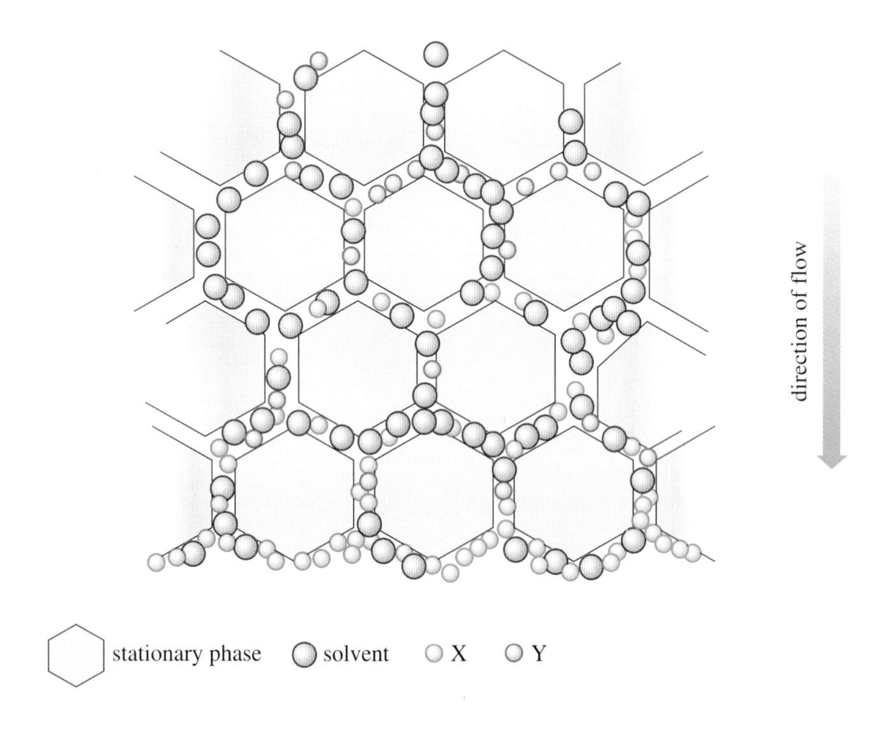

Figure 2.12
A schematic diagram illustrating the column chromatography process. A mixture of molecules X and Y is placed on a column of silica as the stationary phase, where Y is more polar than X and so adsorbs on the column more strongly. As the solvent passes down the column, the molecules of X are eluted more easily and so travel down the column more quickly than Y, thus effecting separation.

High-performance liquid chromatography, HPLC

A more sophisticated method of column chromatography is known as **HPLC, high-performance liquid chromatography**. This employs very fine solid particles which pack closely together. This increases the surface available for adsorption, and so improves the separation, but because the solid particles are packed tightly together, a pump is needed to force the mobile liquid phase through the column.

Gas-liquid chromatography, GLC

Another variant is **GLC, gas-liquid chromatography** *. In this case the mobile phase is a gas, known as the *carrier gas*, and the stationary phase in the column is a liquid — a non-volatile oil or grease. This liquid can be coated on the surface of small particles of an inert solid, which is called *packed-column GLC*, or alternatively it is simply coated on the inside wall of a very long narrow column. In GLC, the point at which the sample is introduced has to be heated so that the components in the mixture vaporize and pass down the column in the gas phase.

* This is also commonly known as gas chromatography (GC).

By slowly heating the column, the time that the less volatile components spend on the surface of the stationary phase can be decreased, and thus the time taken for them to pass along the column (that is, their **retention time**) is reduced.

The essential design of both HPLC and GLC instruments is similar (Figure 2.13). A small amount of the mixture is added into the flowing, mobile phase by injection using a syringe. After passing along the column, the individual components are detected in some way as they come off the column, to give a graphical representation. Figure 2.14 shows a typical output from GLC, known as a chromatogram, where a mixture of hydrocarbons has been separated into its components.

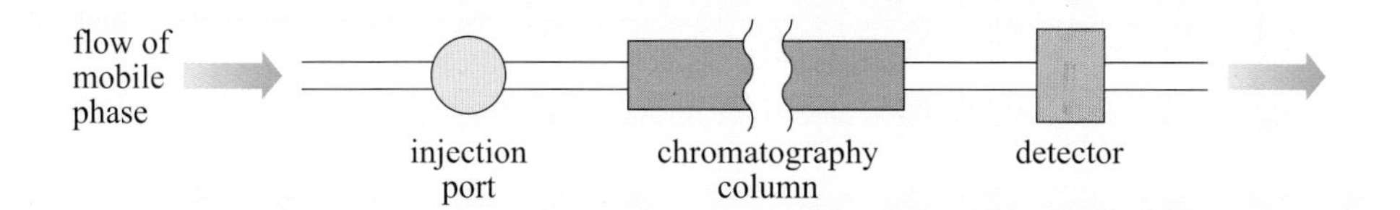

Figure 2.13
The design of HPLC and GLC instruments.

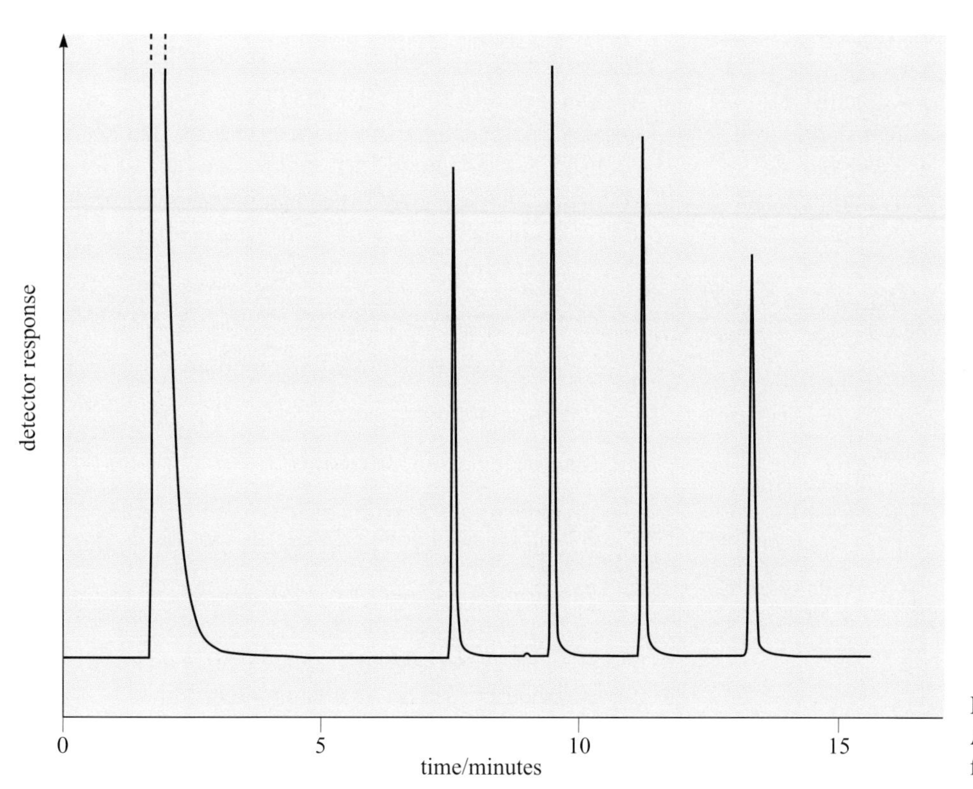

Figure 2.14
An example of a chromatogram from GLC.

COMPUTER ACTIVITY 2.7 Chromatography explained

At some point in the near future you should watch the video entitled *Chromatography explained* in the multimedia activity *Practical techniques* on the *Experimental techniques* CD-ROM that accompanies this book. There you will see an animation of the general chromatographic process, with particular emphasis on GLC. You should also view *GLC* and *HPLC* which show these two techniques in practice. This activity should take approximately 10 minutes to complete.

BOX 2.5 DNA fingerprinting

Deoxyribonucleic acid (DNA) is the genetic material present in the nucleus of every living cell and was first structurally characterized as the now famous double helix by James Watson and Francis Crick in 1951. In humans (and all higher organisms), DNA consists of two strands that wrap around each other to form a double helix (like a twisted ladder), whose sides are made of deoxyribose (a sugar) and phosphate molecules, connected by rungs of nitrogen-containing bases which form hydrogen bonds to each other. Only four bases are present in DNA: *adenine, A, thymine, T, cytosine, C,* and *guanine, G*. The bases bond in pairs: A to T, and G to C are the *only* possible combinations. The order or sequence of the base pairs specifies the exact genetic instructions required to create a particular individual organism, and is slightly different for everyone. Any type of organism may be identified by examination of its DNA sequences. Identifying individuals within a species has proved difficult until recently as this would have meant working out the sequence of 3 billion base pairs. But this process has now been simplified and forms a key tool in the forensic scientist's armoury in identification of suspects. It is known as *DNA fingerprinting* — identifying a person by his or her DNA profile.

The genetic differences between individual persons are due to the differing order of the base pairs in a strand of DNA, but more than 99% is the same for everyone. In 1984 Alec Jeffreys at the University of Leicester noticed that certain segments of a gene, so-called *minisatellites*, didn't do anything, and they took up less than 1% of the total DNA, but they were *different lengths in different people*. He developed a method to extract the DNA from the sample and cut it into segments using specific enzymes. A special type of chromatography, known as *gel electrophoresis**, is then used to separate these fragments on the basis of *size* and *charge*. This technique is similar to chromatography in that it separates on the basis of differing adsorptions of the fragments to a solid support. The difference is that as well as having a mobile phase, an electrical field is also applied. The segments are marked with special radioactive labels and exposed on X-ray film, to give a characteristic pattern of black bars (Figure 2.15). This pattern is then compared with a sample taken from a person or suspect; if the two patterns match, the samples probably came from the same person.

The methods have been developed over the years, and the most recent technique, PCR (polymerase chain reaction) allows the DNA sample to copy repeatedly, so that minute samples can be used for identification. Scientists are now able to use these differences to generate a DNA profile of an individual within hours, using small samples of blood, bone, hair or other body tissue or fluids. Such samples are often obtained at a crime scene.

The first use of DNA fingerprinting in the UK was in 1986 when Jeffreys was asked if he could verify a man's confession to rape. Jeffreys proved that the man was innocent, and the police then looked for a new suspect, taking samples from thousands of men. Colin Pitchfork was eventually arrested for the rape of Lynda Mann and Dawn Ashworth: he had persuaded a friend to give a sample in his stead. Other interesting uses of DNA fingerprinting and identification include:

- Miscarriages of justice;
- Identifying lost children in Argentina;
- Paternity/maternity cases;
- Identifying accident victims;
- Identifying the remains of murdered Nicholas Romanov, the last Tsar of Russia and his family;
- Confirming the identity of the son of Louis XVI and Marie-Antoinette. Royalists argued for 205 years over whether Louis-Charles de France perished in 1795 in a Paris prison or rather escaped the French Revolution. The heart of the child was tested in 1999 for its DNA and compared with the DNA from a lock of hair of Marie-Antoinette. They were compatible and this proved that her son died in prison.
- Migration patterns: bioanthropologists have used DNA fingerprinting to trace human migration patterns across the world

* Gel electrophoresis is explained in the Case Study, *Polymers* in *Mechanism and Synthesis*[3].

BOX 2.6 The Human Genome Project

The Human Genome Project shows how molecular separation and identification methods can work in tandem.

The complete set of instructions for making an organism is called its *genome*; it is the master blueprint for all cellular structures and activities for the lifetime of the cell or organism. Found in the cell nuclei of every person's billions of cells, the human genome consists of tightly coiled threads of DNA and associated protein molecules, organized into structures called chromosomes. The plan to determine the complete human genome sequence was outlined by an international consortium in 1995 and the results were published in 2001. DNA sequencing — the determination of the exact order of the 3×10^9 bases making up the DNA of the 23 different human chromosome pairs — is the hardest task of the human genome project.

Although far more complex than we can consider here, the process essentially involves just two chemical methods, which we have already discussed — separation and identification. After the chromosomes are broken down into many shorter pieces, the fragments are first separated using gel electrophoresis. Each separated fragment is then identified, often employing mass spectrometry (see Section 5.3) as the key technique.

Figure 2.15 A DNA fingerprint.

2.4 Recrystallization

Solid substances are often purified by a procedure known as **recrystallization**. The impure solid is dissolved in the minimum amount of an appropriate hot solvent and, after filtration of the hot solution to remove any insoluble impurities, the solution is allowed to cool. Crystals of the required substance start to crystallize out, leaving the soluble impurities in solution. When cool, the crystals can be filtered off.

COMPUTER ACTIVITY 2.8 Recrystallization

At some point in the near future you should watch the video entitled *Recrystallization* in the multimedia activity *Practical techniques* on the *Experimental techniques* CD-ROM that accompanies this book. The sequence shows this technique in practice, and discusses the theory of the procedure. The product that is recrystallized in this sequence is benzoic acid, which was obtained by the oxidation of benzyl alcohol. This activity should take approximately 10 minutes to complete.

In this book, we have talked about filtration a great deal — such as filtering off a drying reagent like magnesium sulfate or filtering a hot solution during a recrystallization. You should be familiar with filtration from everyday life: for example, every time you drain some vegetables or pasta in a colander, you are effectively performing a filtration! The solid (food) remains in the colander while the water drips through and away (Figure 2.16). It is the same with a filtration in the laboratory — the solid (crystals) remains in the funnel on the filter paper while the liquid runs off and into a collection flask. There are several different methods for the filtration of a material; the two primary methods being filtration under gravity and filtration under suction. You should recall the contrasting appearance of the samples of suction-filtered and gravity-filtered crystals of benzoic acid.

Figure 2.16 'Filtering' peas in a colander.

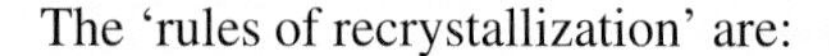

The 'rules of recrystallization' are:

1. If you want the *solution* (such as in hot recrystallization), filter under gravity through a fluted (highly folded) filter paper in a glass funnel.
2. If you want the *solid*, filter under suction through a flat filter paper in a porcelain (Buchner or Hirsch) funnel.

Recrystallization is an excellent method for the purification of solids and is the last of the major purification techniques. It is used by both inorganic and organic chemists. The preparation of many inorganic salts and complexes leads to the formation of solid crystals, which can be purified by recrystallization and filtration.

For example, transition metal complexes such as $[Co(NH_3)_5Cl]Cl_2$ and $[Co(NH_3)_6](NO_3)_3$, main-group species like $[PCl_4][SbCl_6]$, and the inorganic heterocyclic compounds

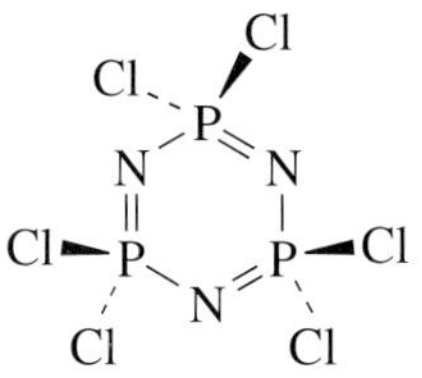

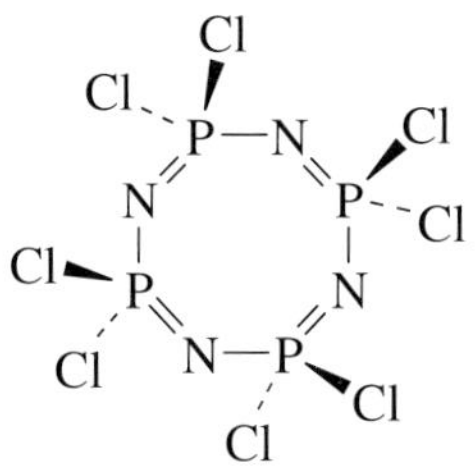

are all inorganic compounds that are typically purified by crystallization and filtration with appropriate solvents.

Before we move on to the next step in preparation, let's first summarize our separation and purification techniques and consider when to use each one.

2.5 Which technique to use?

For a particular mixture of substances, how do we go about selecting a method of separation? The technique chosen depends very much on the type of problem we're faced with. You have probably realized that the technique of recrystallization is applied to both organic and inorganic compounds, since many of the former and most of the latter, are solids. By a similar argument, you should understand that distillation can only be applied to liquids and tends to be used mostly during organic preparations. Any material that may be dissolved in an organic solvent (irrespective of polarity) may be subjected to chromatographic techniques and so this applies equally to organic and inorganic materials.

QUESTION 2.8

Fill in Table 2.3, giving three ticks for a technique that completely satisfies the criterion on the left, down to one tick where the fulfilment is poor.

Table 2.3 The choice of techniques for separation

	Distillation	Solvent Extraction	Chromatography
1 Will separate a small number of components in a mixture			
2 Will separate a large number of components			
3 Will separate small amounts of material			
4 Will separate large amounts of material (i.e. on an industrial scale)			

The important points to note are that for the separation of small amounts of a complex mixture, chromatography is supreme. The technique is simple to operate, comparatively cheap to run, and is quick, especially in the case of gas-liquid chromatography (GLC). As the amount of material increases, or the complexity of the mixture decreases, the other techniques increase in applicability and become most powerful as methods for the industrial separation of simple mixtures. Distillation and solvent extraction at the industrial level are fairly comparable in speed, cost and simplicity.

As we shall see, in order to identify substances, we utilize techniques that require only very small amounts of the pure substances. Thus, if we are presented with the task of identifying the components of an organic mixture, we need to take only a small sample, so the technique to choose to separate these components is invariably chromatography.

2.6 Summary of Section 2

1 Performing a chemical reaction is often easier than the reaction work-up, purification and characterization of the reaction products.

2 The techniques of separation and purification most commonly employed are solvent extraction, distillation, chromatography and recrystallization. More than one of these techniques may be needed.

3 Solvent extraction takes advantage of differences in solubility in particular solvents. The separation achieved is somewhat crude, but the technique can cope with large amounts of material.

4 Distillation is appropriate for separating large amounts of a liquid mixture containing a small number of components with large differences in boiling temperature.

5 The best separation of components is provided by the chromatographic techniques, but TLC, GLC and HPLC can separate only relatively small amounts of material. These techniques are ideal for separating the components of a mixture prior to molecular identification. Column chromatography is useful for handling larger amounts of mixtures, but the separation is less efficient than HPLC or GLC.

QUESTION 2.9

Which technique would you employ to purify large amounts of an organic liquid?

QUESTION 2.10

Which technique would you employ to separate small amounts of benzene (boiling temperature 80.1 °C) and cyclohexane (boiling temperature 80.8 °C)?

3 COMPLETING A SYNTHESIS

Once you have completed a reaction, separated and purified the product(s), there are still questions to be answered:

- How much compound have you made?
- Are you sure it is pure?
- What is the formula?
- What is the structure?

The first question is addressed by working out the percentage yield *, and Table 3.1 gives an idea of the expectations here. Section 4 addresses the problems of checking purity and Section 5 considers identifying the new compound. Determining the structure of a compound is a very complex process, and we devote all of Part 2 of this book, which is to be found on the *Spectroscopy* CD-ROM, to answering this question.

Table 3.1 Reaction yields

% yield	Comment
0	Reaction failed!
0 to 20	Often called a 'formation' rather than a 'synthesis', the desired product is formed, but in yield too low to be of any practical use in a commercial synthesis.
20 to 40	Modest yield
40 to 70	Good yield
>70	Excellent yield

Chemists, particularly when considering a reaction for a commercial purpose, are only interested in yields in the last two categories. In research and development work, smaller yields are acceptable because only small quantities of reactants are usually used.

* Percentage yield is discussed in *Alkenes and Aromatics*[4].

CHECKING FOR PURITY

4

4.1 How pure is pure?

It is easy to assume that a substance you have prepared is pure — that it contains no substances other than the one of interest. However, it is always the case that, on preparing a compound, the initially isolated product is the required substance contaminated with small amounts of by-products and starting materials.

We cannot claim absolute purity for any substance because of the limitations of the techniques used for detecting the presence of impurities. In other words, purity can only be guaranteed to within the limits of detection. Nevertheless, if the level of purity is acceptable for our purposes, then we can consider the substance as pure. The purity required of the product depends on what it is needed for. Take the example of tap water: when we drink it, we take its purity for granted. However, drinking water is far too impure to use to top up a car battery or a steam iron: it contains antibacterial agents such as chlorine, and also calcium and magnesium salts which have been dissolved during the percolation of rain water through the soil into the reservoirs and rivers. When pure water is needed we used deionized or distilled water.

Purity is usually expressed as a percentage, e.g. 99.9% pure. Chemicals for experiments can normally be purchased in two grades. The *standard laboratory reagents* will be about 99% pure, sometimes less, depending on the chemical and how it has been made and purified. The impurities and their concentrations are listed on the bottle. The *analytical reagents*, when available, will be of the order of 99.9 or 99.99% pure: they are much more expensive, and may only be used for specialist purposes such as spectroscopy and analysis.

A final drug substance, which will be sold by the pharmaceutical industry, must be free of even the smallest trace impurities. The chemist must remove the impurities, using the various separation techniques that we have already discussed in the previous section.

- Once a technique has been used to purify a product, what means do we have of testing the purity?

- You might suggest a number of possibilities: chromatography can tell us how many different components are present. Melting temperatures and boiling temperatures can be determined as they are sensitive to purity.

We have already seen examples, when in Computer Activity 2.4 TLC was used to separate several components, and in Computer Activity 2.7 GLC was employed to examine the spirit distilling in whisky production. The 2-chloro-2-methylpropane product of the preparation discussed in Sections 1 and 2 is known to boil at 49 °C to 51 °C, so if the product of the reaction boils in the same narrow temperature range, it is likely to be pure 2-chloro-2-methylpropane.

Many solids have very high boiling temperatures, so it is more appropriate to measure their melting temperatures than their boiling temperatures, and this is an easy and quick measurement to make. However, whether it be melting or boiling temperature, the principle is much the same in either case: if a solid is pure, it will melt sharply (that is, over a narrow range of temperature) and a pure liquid will have a well-defined boiling temperature.

COMPUTER ACTIVITY 4.1 Measuring a Melting Temperature

At some point in the near future you should watch the video entitled *Measuring a Melting Temperature* in the multimedia activity *Practical techniques* on the *Experimental techniques* CD-ROM that accompanies this book. The sequence shows how to measure the melting temperature of a solid. This activity should take approximately 10 minutes to complete.

Samples can be claimed to be pure only to the extent that, for all practical purposes and with currently available techniques, no impurity is detectable.

QUESTION 4.1

Which technique(s) would you employ to check the purity of a commercial sample of (i) an organic solid, (ii) an organic liquid?

5 IDENTIFYING A COMPOUND

Thus far we have seen how to plan and perform a reaction, to separate and isolate the various products, and check for purity. Sometimes we know what the product or products of a reaction are likely to be, for example, if we are repeating a procedure which is already published in a book or scientific paper. On other occasions, as is the case in research, we know what we hope the product will be, but we cannot be certain. So what do we do in these cases? There is no advantage in obtaining a pure, but unknown, product from a reaction, we have to be able to *characterize* it. The identification techniques available are mostly applicable to both organic and inorganic chemistry, but the priority that each branch of chemistry gives to each technique tends to vary.

If the preparation has been previously published in a book or scientific paper, then it is usually possible to identify it by comparison with various data listed in the literature. However, the compound may be completely unknown, or maybe it has never been made before so there is no data for comparison, or we may have an unexpected product. We shall see that the methods that are of most use to us in identification through comparison, also allow us to identify molecules when no comparison is possible; that is, they allow us to infer the molecular identity.

Many identification procedures that we use for previously recorded compounds are based on the measurement of some physical property of the compound. We will already have noted some physical properties when the separation was carried out. For example, if we used distillation, we would have noted the boiling temperature; if we used chromatography, we would have noted the time taken for the component to travel a certain distance (or the R_f value); or in the case of solvent extraction, we would have noted the solubility characteristics in various solvents. There are many other measurements that we can make. We could determine the mass of the molecule (relative molecular mass), the density of the substance, the acidity, or the amount and frequency of electromagnetic radiation absorbed.

There is therefore a range of analytical techniques we can employ, but we need to decide which would help most towards the identification of a particular compound. We will start by looking at the techniques which identify the different elements, and which measure how much of each element is present. This allows the formula of the compound to be determined. These techniques are expensive and *destructive* (you cannot recover your sample) so they tend to be used only as a final check before the publication of results.

5.1 Elemental analysis

5.1.1 Carbon, hydrogen and nitrogen analysis

For inorganic and organic compounds the most common analysis undertaken is the determination of the amounts of carbon, hydrogen and nitrogen present. The usual method used is known as **combustion analysis**, where an accurately weighed amount of the compound is burnt in oxygen to form CO_2, H_2O, and N_2 respectively.

These are then selectively collected on weighed adsorbents, and the increase in mass of the adsorbent is determined. Note, however, that combustion analysis is a destructive technique, and can be expensive, and so it is often not employed until other characterization has taken place, and only then if you are publishing your work and have to prove beyond doubt that you can substantiate your claims.

COMPUTER ACTIVITY 5.1 Combustion Analysis

At some point in the near future you should watch the video entitled *Combustion Analysis* in the multimedia activity *Practical techniques* on the *Experimental techniques* CD-ROM that accompanies this book. The sequence shows the operation of a laboratory elemental analyser for C, H and N. This activity should take approximately 5 minutes to complete.

5.1.2 Other elemental analyses

A number of other common elements can be analysed using similar principles to the C, H and N analysis. In fact the only element which is *not* easily determined is *oxygen* — the oxygen content is usually inferred as a residual mass from the quantities of the other elements.

Sulfur

The sulfur in organic and biological materials is determined by burning in a stream of oxygen. The SO_2 produced is reacted with hydrogen peroxide to form sulfuric acid, which can then be titrated *

$$S + O_2(g) \longrightarrow SO_2(g)$$

$$SO_2(g) + H_2O_2(aq) \longrightarrow H_2SO_4(aq)$$

Nitrogen

An alternative to combustion analysis is the *Kjeldahl method*; the nitrogen-containing sample is decomposed in hot concentrated sulfuric acid which converts the bound nitrogen to the ammonium ion. The solution is cooled, diluted and made basic to release ammonia. The released ammonia is collected and titrated. This is the standard method for the determination of the protein content of grains and meats, as multiplication of the percentage of nitrogen by a suitable factor (6.25 for meats and 5.7 for cereal) will give the percentage of protein in the sample. Table 5.1 summarizes a few of these methods.

Table 5.1 Summary of some elemental analysis methods

Element	Converted to	Absorption Products	Titration
N	NH_3	$NH_3 + H_3O^+ \longrightarrow NH_4^+ + H_2O$	Excess HCl with NaOH
S	SO_2	$SO_2 + H_2O_2 \longrightarrow H_2SO_4$	NaOH
C	CO_2	$CO_2 + Ba(OH)_2 \longrightarrow BaCO_3 + H_2O$	Excess $Ba(OH)_2$ with HCl
Cl (Br)	HCl	$HCl + H_2O \longrightarrow Cl^- + H_3O^+$	NaOH
F	SiF_4	$SiF_4 + H_2O \longrightarrow H_2SiF_6$	NaOH

* Titration is discussed in *Exploring the Molecular World*[1].

5.1.3 Atomic spectroscopy

Atomic spectroscopy in various forms, is used for the qualitative and quantitative analysis of about 70 elements. First, we revise the principles behind the technique.

Each free uncombined atom or ion of a chemical element has a set of electronic energy levels characteristic of that element. The electrons occupy the levels of lowest energy, when the atoms are said to be in their *electronic ground state*. The electrons can however be excited to higher energy levels, when the atoms are said to be in an *excited state*. In order to reach the excited state, the atom or ion must take up the exact amount of energy to transfer it to the excited state — a process known as *absorption* (Figure 5.1). Conversely if the atom is in an excited state, it can return directly to the ground state when it gives out this same amount of energy as a photon of radiation, a process called *emission* * (Figure 5.1).

The energy difference between the two energy levels, ΔE is related to the frequency of the light that is emitted, ν, by the Einstein relation

$$\Delta E = h\nu \qquad (5.1)$$

where h is Planck's constant and has the value $h = 6.626 \times 10^{-34}$ J s. In spectroscopy, the frequency of radiation is usually denoted by the Greek letter nu, ν.

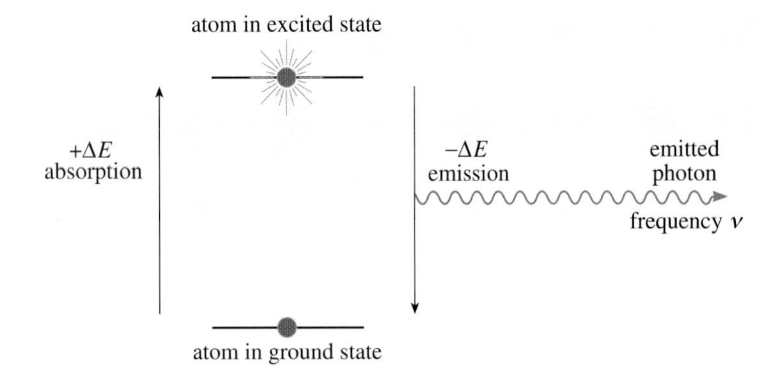

Figure 5.1
The absorption of energy and subsequent emission of radiation by an atom.

You will be familiar with this phenomenon from the atomic spectrum of the hydrogen atom, which is shown with that of some other elements in Figure 5.2, and in the characteristic flame tests for some of the metals, which emit bright colours when vaporized in a hot flame (Figure 5.3). If you need to refresh your memory, this topic is revised in the first section of the *Spectroscopy* CD-ROM, '*Introduction to Spectroscopy*'.

COMPUTER ACTIVITY 5.2 An Introduction to Spectroscopy

At some point in the near future you should study *An Introduction to Spectroscopy* on the *Spectroscopy* CD-ROM that accompanies this book. The principles of atomic spectroscopy are revised in this sequence. This activity should take approximately 1.5 hours to complete.

---

* There is another process by which an excited species can return to the ground state, known as *fluorescence*, which is not discussed in this book.

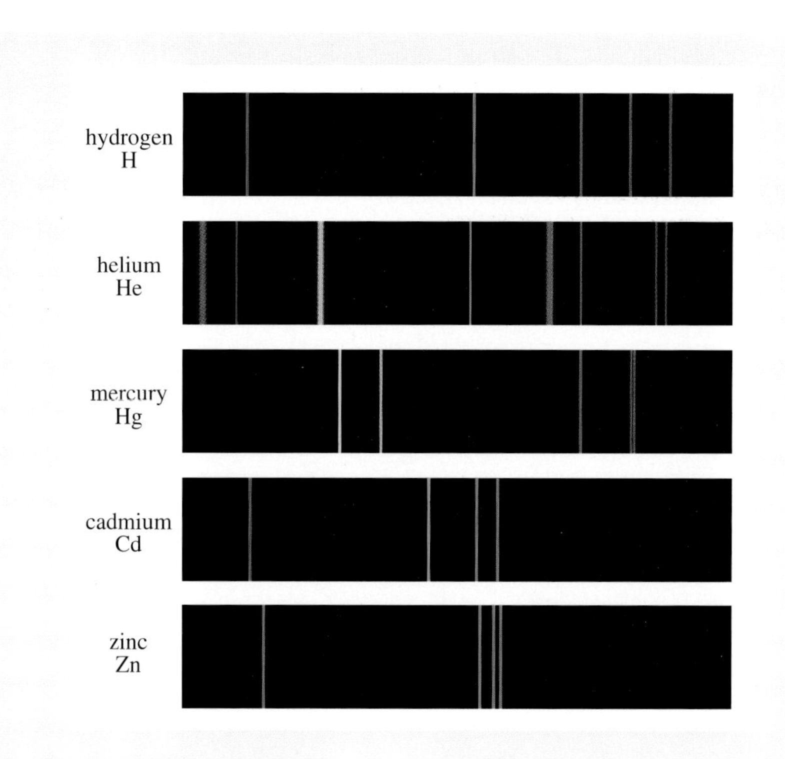

Figure 5.2
The emission spectra of hydrogen, helium, mercury, cadmium and zinc in the visible region of the electromagnetic spectrum.

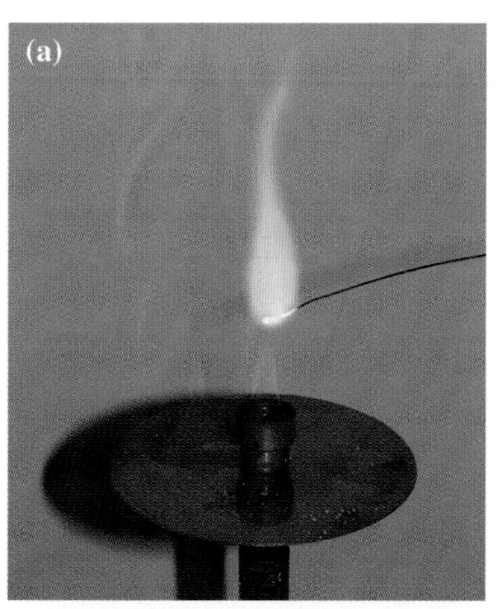

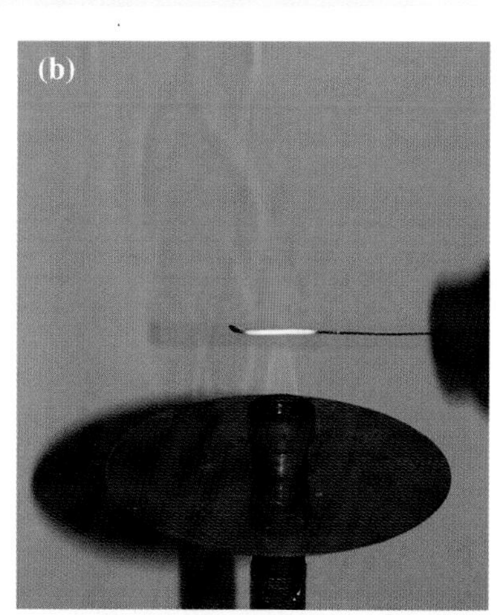

Figure 5.3
Characteristic colours produced in flame tests for the metals
(a) sodium
(b) strontium
(c) barium
(d) potassium
(e) calcium

Each atom or ion has a characteristic set of electronic energy levels, and so either absorbs or emits a particular set of frequencies in its atomic spectrum. By measuring these frequencies and their intensity we are able to identify a particular element and measure the amount present. Figure 5.4 shows some of the energy levels for sodium, and the characteristic absorption pattern that they would produce. The full spectrum of sodium consists of about 40 peaks, but for elements that have several outer electrons that can be excited, the spectrum can be very complex, sometimes consisting of hundreds of peaks.

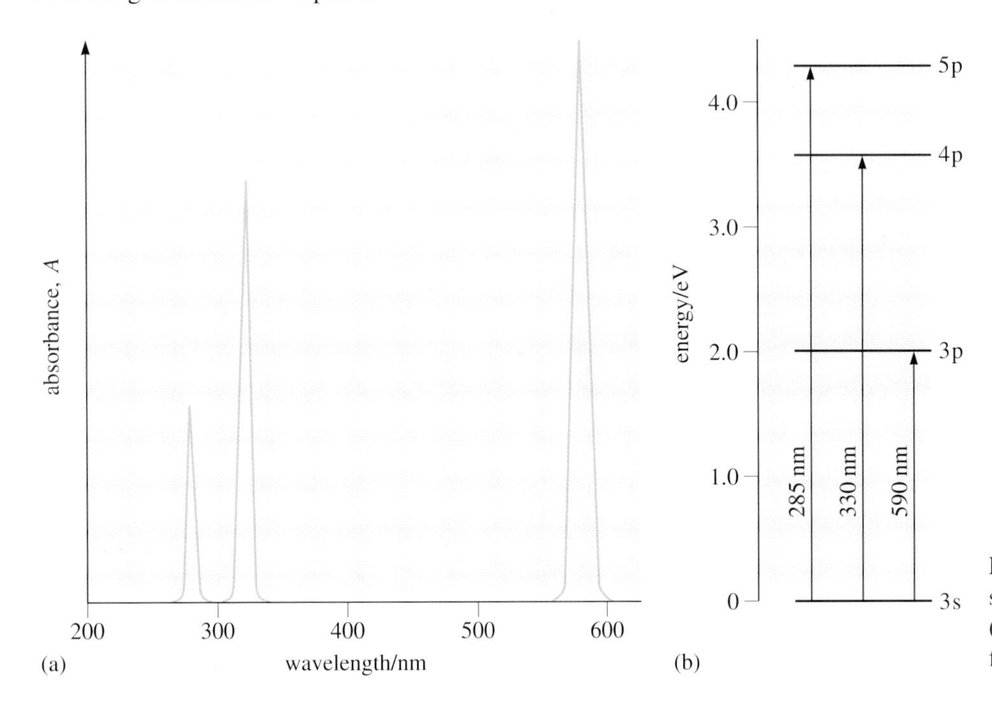

Figure 5.4 (a) Partial absorption spectrum for sodium vapour. (b) Electronic transitions responsible for the lines in (a).

QUESTION 5.1

The excited state of sodium which gives rise to a characteristic orange–yellow light, lies 2.1 eV above the ground state. (i) What is the frequency of the orange light? (ii) Determine the wavelength of the orange light and thus what transition it is due to. ($1\ \text{eV} \equiv 1.602 \times 10^{-19}\ \text{J}$) Take any other data from the *Data Book* (available on the CD-ROM).

Spectroscopic determination of atomic species can only be carried out in the gas phase, where the individual atoms or ions are well separated. Consequently, the first step in the process is *atomization*, where the sample is volatilized (heated to the gas phase) and decomposed to produce an atomic gas. The differences between the various atomic spectroscopy techniques available, largely lie in the different ways of doing this. The most widely used method is *flame atomization*, where the sample is decomposed in a flame (a sophisticated version of the common flame test), but other common methods (Table 5.2) are

- electrothermal,
- inductively coupled plasma,
- direct-current plasma.

We look below at how electrothermal atomic absorption spectroscopy works in some detail.

Electrothermal atomizers offer unusually high sensitivity for small volumes of sample because the whole sample is atomized in a very short time. Figure 5.5 shows a particular set-up used for the determination of the lead content in a food sample, such as apple juice.

The centrepiece is the electrothermal atomizer. It consists of a graphite platform seated inside a graphite tube through which a large electric current can be passed. A small sample is delivered by microsyringe through the hole in the top. The electric current is used to heat the sample in three stages. In the first stage at about 150 °C the sample is dried by the evaporation of water; in the second stage, at about 600 °C, it loses any volatile organic matter and is charred to an ash; in the third stage, the temperature is raised to 2 000 °C, when the lead and other elements are vaporized and converted to free atoms.

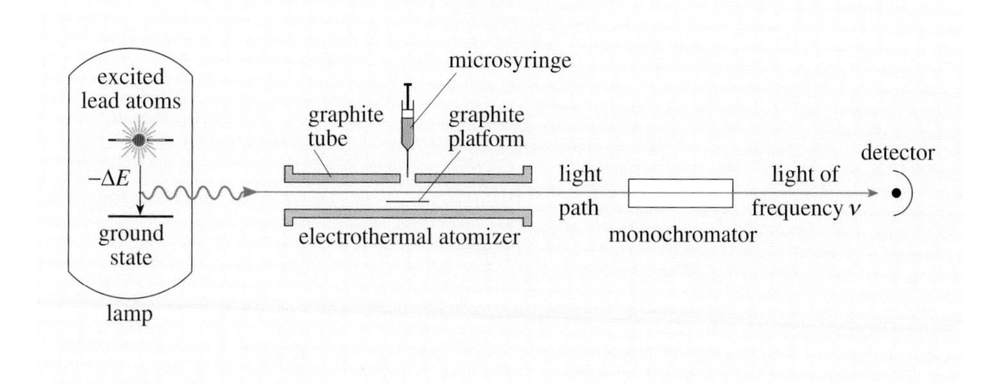

Figure 5.5
An atomic absorption spectrometer using electrothermal atomization.

The lead-bearing vapour lies in the path of radiation arising from a source of lead atoms in excited electronic states. The radiation is produced in a special lamp, by bombarding a target made of metallic lead with fast-moving noble gas positive ions such as neon, Ne^+. The excited lead atoms will quickly revert to the ground state, and as they do so, emit radiation of characteristic frequency, ν.

If there is no lead-containing vapour above the graphite platform, the light will pass through the tube without losing any intensity.

- What happens if there is a vapour containing lead atoms above the platform?

- If there is a lead-containing vapour present, then some of the characteristic radiation will be absorbed by the lead atoms, thus reducing the intensity of the radiation.

The amount of light absorbed is usually plotted out as the **absorbance**, A, which is proportional to the lead concentration of the sample (Figure 5.6)

$$A = \varepsilon c l \quad (5.2)$$

where c is the molar concentration of the absorbing species, l is the *pathlength* (the distance the light travels through the sample) and ε is a proportionality constant, known as the *molar absorption (extinction) coefficient*. The absorbance is calculated by measuring the intensity of the incident light on the sample, I_0, and the intensity of the light after it has travelled through the sample, I:*

$$A = \log_{10}\left(\frac{I_0}{I}\right) \quad (5.3)$$

* The ratio I/I_0 is known as the *transmittance*.

On the left of Figure 5.6, the first three signals record the absorbances of three 'standards' with known concentrations of lead. These were solutions of lead nitrate in water containing 0.05, 0.1 and 0.2 micrograms of lead per millilitre of solution ($\mu g\ ml^{-1}$). Samples of these solutions with a volume of 2 microlitres (2 µl) were dispersed in turn on to the platform of the electrothermal atomizer.

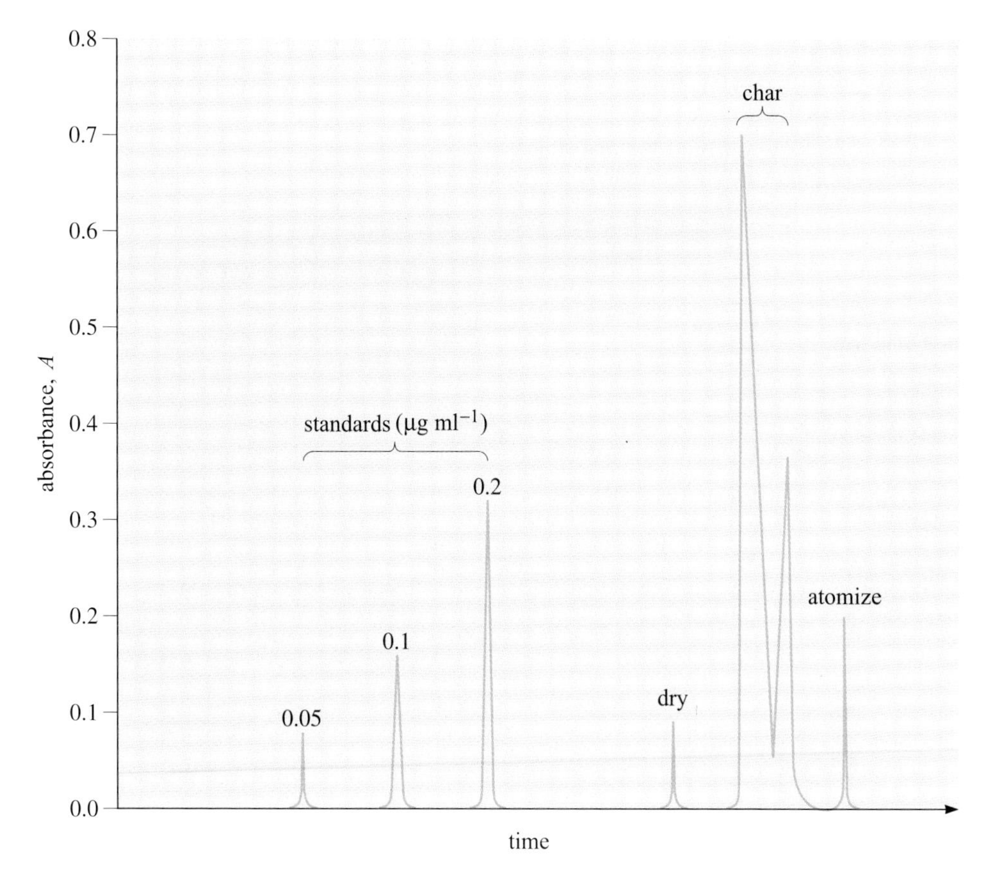

Figure 5.6 Typical output from an atomic absorption spectrometer fitted with an electrothermal atomizer, which is being used to determine the concentration of lead in apple juice. Absorbance is plotted on the vertical axis.

- How do the three signals bear out our claim that the absorbance, *A* of a sample is proportional to its concentration?

- The absorbances (measured as the peak areas, which are proportional to the peak heights) are in the same ratios as the concentrations. Thus the third peak is twice as high as the second, and the second peak is twice as high as the first.

The remaining peaks record the absorbance of a 2 µl (= 2×10^{-6} litres) sample of apple juice. In this case, the sample also contains organic materials, and these give rise to vapours which absorb radiation of the chosen frequency during both the drying and the charring stages; this is the origin of the peak marked 'dry' and the two peaks marked 'char'. Only during the last stage, atomization, do the lead atoms get into the light beam, so the final peak on the right is the one that yields the lead concentration.

- Use this peak and that for the standard of concentration 0.2 $\mu g\ ml^{-1}$ to calculate the concentration of lead in the apple juice.
- Because absorbance is proportional to concentration

$$\frac{\text{concentration of lead in juice}}{\text{concentration of lead in standard}} = \frac{\text{absorbance for juice}}{\text{absorbance for standard}}$$

Now the absorbance of the apple juice is the height of the peak marked 'atomize' when read off on the vertical axis; this is 0.20. Likewise, for the standard of concentration 0.2 $\mu g\ ml^{-1}$, this height or absorbance is 0.32. Thus

$$\frac{\text{concentration of lead in juice}}{0.2\,\mu g\ ml^{-1}} = \frac{0.20}{0.32}$$

Therefore

$$\text{concentration of lead in juice} = \frac{0.20}{0.32} \times 0.2\,\mu g\ ml^{-1}$$
$$= 0.125\,\mu g\ ml^{-1}$$

Table 5.2 Some of the atomic spectroscopy techniques commonly available and their acronyms

Atomization method	Typical atomization temperature/°C	Spectroscopic basis	Common name and abbreviation
Flame	1 700 to 3 150	Absorption	Atomic absorption spectroscopy, AAS
		Emission	Atomic emission spectroscopy, AES
Electrothermal	1 200 to 3 000	Absorption	Electrothermal atomic absorption spectroscopy
Inductively coupled argon plasma	6 000 to 8 000	Emission	Inductively coupled plasma spectroscopy, ICP
Direct-current argon plasma	6 000 to 10 000	Emission	Direct-current plasma spectroscopy, DCP
Electric arc	4 000 to 5 000	Emission	Arc-source emission spectroscopy
Electric spark	40 000	Emission	Spark-source emission spectroscopy

5.2 Finding the empirical formula

If we take any hydrocarbon and burn it completely in oxygen, the products of the reaction are CO_2 and H_2O. For example, if 56 g (one mole) of C_4H_8 were burnt completely in oxygen, 176 g of CO_2 and 72 g of H_2O would be produced, in agreement with the balanced equations

$$C_4H_8 + 6O_2 = 4CO_2 + 4H_2O$$
$$56\,g + 192\,g = 176\,g + 72\,g$$

So, can we take a known mass of a hydrocarbon, weigh the products of the combustion reaction and hence determine the **molecular formula** of the hydrocarbon, C_4H_8 in this case? Let's consider combusting the above amount of substance (although in practice milligram amounts are used). We take 56 g of the hydrocarbon C_xH_y, burn it, and the result is 176 g of CO_2 and 72 g of H_2O. Can we determine the values of x and y? Unfortunately we can't because burning 56 g of C_2H_4, C_3H_6, C_5H_{10}, or indeed $C_{37}H_{74}$, would also produce exactly the same amounts of products. Let's check that this is so.

Write the equation for the burning of C_2H_4 in oxygen.

$C_2H_4 + 3O_2 = 2CO_2 + 2H_2O$

According to this equation, the molar mass of C_2H_4 (28 g) would give twice the molar masses of carbon dioxide ($2 \times 44 = 88$ g) and water ($2 \times 18 = 36$ g). If 28 g of C_2H_4 give 88 g of CO_2 and 36 g of H_2O, then starting with double the amount of C_2H_4 (56 g) would give double the quantities of the products, that is, 176 g of CO_2 and 72 g of H_2O, that is, the same amounts generated from 56 g of C_4H_8. If you still need convincing, check that 56 g of C_3H_6, C_5H_{10} and $C_{37}H_{74}$ give exactly the same results.

Why do all these different compounds give the same results?

Because they all have the same *empirical formula* namely, CH_2.

So, although we cannot determine the values of x and y in the hydrocarbon, what we can do is determine the percentage by mass of a particular element in the compound, and thus the empirical formula. Now have a go at another example.

QUESTION 5.2

0.023 4 g of an organic compound produced 0.079 2 g of carbon dioxide and 0.016 2 g of water on combustion analysis. Given that the relative atomic masses of carbon, hydrogen and oxygen are 12, 1 and 16 respectively, calculate the percentage carbon, hydrogen and oxygen present in the compound.

The oxygen content of a compound is not usually detected by elemental analysis, but is normally determined by difference, i.e. the remainder when all the other percentages have been subtracted from 100.

Once all the percentages are known, it is possible to calculate the empirical formula of the compound by converting *mass ratios* into *atomic ratios*. The method is illustrated with the following example.

Example 5.1

You may receive the following data: an organic compound contains 40% carbon and 6.7% hydrogen by mass.

Step 1: Find the percentage by mass of oxygen in the compound (by difference). In this example

$$100 - (40 + 6.7) = 53.3\%.$$

Step 2: Assume you have 100 g of the compound, find the number of moles of each element present; this gives you the atomic ratios.

To do this you divide the number of grams of each element by the relative atomic mass of the element. In this example there would be

$$\frac{40}{12} = 3.33 \text{ moles of C; } \frac{6.7}{1} = 6.7 \text{ moles of H; and } \frac{53.3}{16} = 3.33 \text{ moles of O}$$

Thus the atomic ratios are

C : H : O = 3.33 : 6.7 : 3.33

Any other elements such as metals in an inorganic compound are treated in exactly the same way — find the number of moles by dividing the number of grams by the relative atomic mass in grams.

Step 3: Divide through by the lowest atomic ratio to give whole numbers; this gives the empirical formula. In this example the lowest atomic ratio is 3.33, so we divide through by 3.33 to give the atomic ratios,

C : H : O = 1 : 2 : 1

giving the empirical formula CH_2O.

It can be helpful to tabulate the calculations.

Table 5.3 Summary of the calculation in Example 5.1

Element	C	H	O
Percentage by mass	40	6.7	53.3
Number of moles	$\frac{40}{12}$	$\frac{6.7}{1}$	$\frac{53.3}{16}$
Atom ratios	3.33	6.7	3.33
Dividing through by the lowest atomic ratio	$\frac{3.33}{3.33}$	$\frac{6.7}{3.33}$	$\frac{3.33}{3.33}$
Whole number ratio of atoms	1	2	1

Thus from the percentage masses of the elements, it is possible to determine the empirical molecular formula. If we were expecting to prepare a compound with an empirical formula of CH_2O, then these results indicate that we may have the correct compound. If we had inadvertently prepared the wrong compound, then this formula would not fit.

Try Question 5.3 now. There are more questions at the end of the section.

QUESTION 5.3

Calculate the empirical formula of an organic compound containing 81.8% carbon and 18.2% hydrogen.

The only problem as far as structural determination is concerned is that elemental analysis only gives us the empirical formula and not the molecular formula. That is where another very important technique comes in — mass spectrometry.

5.3 Mass spectrometry

Mass spectrometry can be used to separate the different isotopes of a single element or can be used to fragment a molecule and separate out the different molecular fragments – a technique used to help in the identification of a molecule.

Natural lead contains four isotopes ^{204}Pb, ^{206}Pb, ^{207}Pb and ^{208}Pb. In a mass spectrometer a lead compound is heated by a *thermal ion emission source*, which consists of a tungsten metal filament in a high vacuum. This breaks down the lead compound and creates individual positively charged lead ions, Pb^+. These ions are accelerated by a high voltage, and at the same time subjected to a strong magnetic field at right angles to their direction of motion (Figure 5.7). The charged species are deflected by the magnetic field. Since each species carries a single positive charge, the amount of deflection in the magnetic field is dependent solely on the mass of the ion. Ions of high mass are deflected to a smaller extent than those of a lower mass, so the ions $^{204}Pb^+$, $^{206}Pb^+$, $^{207}Pb^+$ and $^{208}Pb^+$ separate into four different trajectories. The four ions are collected separately and recorded as a **mass spectrum**. The amount of deflection allows the mass of the ion species to be calculated. The separated species can each be collected by an electronic measuring device. The size of the electric current measured by this device tells us the abundance of the species. Figure 5.8 shows a mass spectrum for a lead sample; there is a peak for each isotope, and the peak heights are proportional to the abundances.

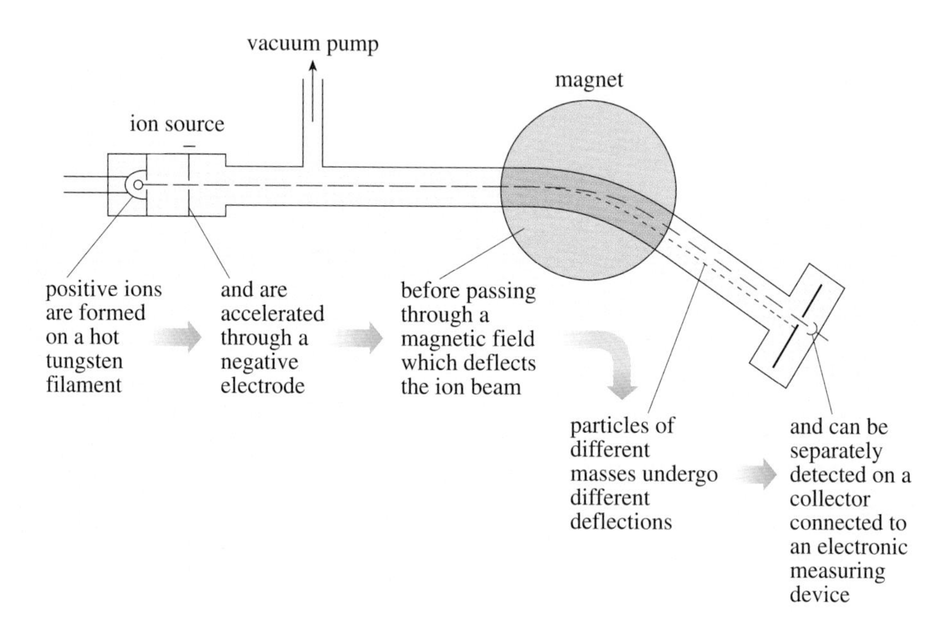

Figure 5.7 The principles and components of a mass spectrometer.

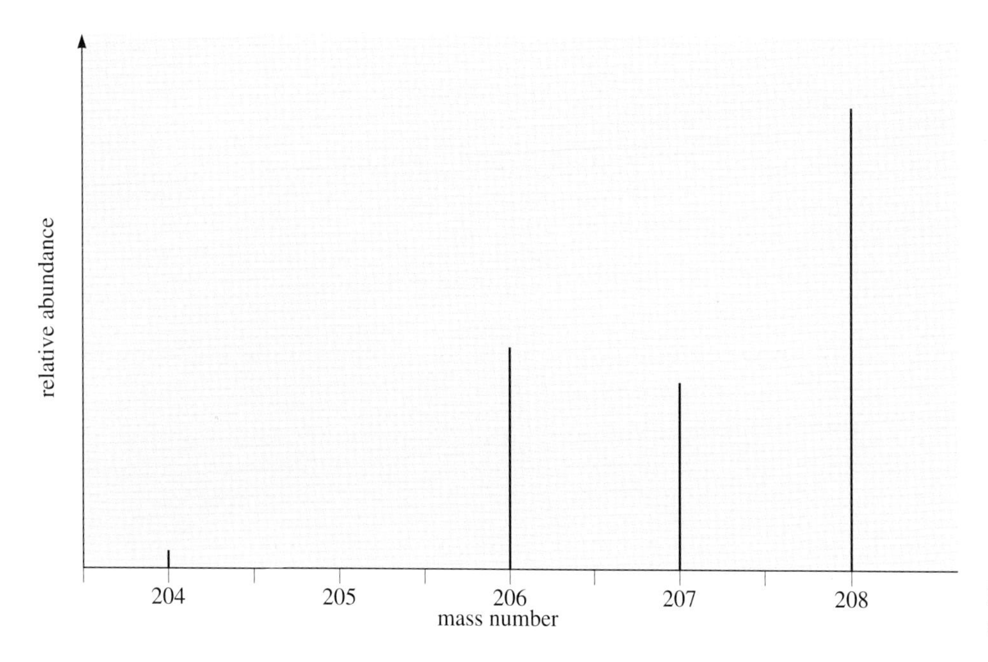

Figure 5.8 The mass spectrum of a sample of naturally occurring lead.

QUESTION 5.4

What are the most common and least common isotopes of naturally occurring lead? How many atoms of ^{208}Pb are there for every atom of ^{207}Pb?

BOX 5.1 Lead in the Arctic ice layers

Analysing the Arctic ice layers for lead can give information on how the global use of lead has changed over the ages.

The climate in the extreme polar regions of the world is very dry. The annual layers of snow are usually quite thin, and because of the constant extreme cold, there is little mixing of the snow layers from one year to another. Thus the age of the snows increases with depth and can be dated. Analysing different sections of a core of ice taken down through the layers will give us information on how air pollution has changed over the centuries. When analysing for lead, the amounts of lead to be measured are extremely small and special techniques have to be employed. Dr Clair Patterson (Figure 5.9) has used *isotope dilution analysis* to study the lead concentrations of the differently aged layers in Northern Greenland.

Figure 5.9 Clair Patterson of the California Institute of Technology. His results initiated the campaign against environmental lead, which finally led to the ban on leaded petrol in the UK in 2000.

Consider an ice sample with a mass of about 20 kg, it may contain as little as 0.2 μg of lead. This means that when the ice melts, this amount of lead will be dispersed in about 20 litres of water. How can such a small concentration be measured? Lead is usually analysed by precipitating PbS from the solution and weighing it. In the case of very dilute solutions though, the lead concentration to begin with is so low that the proportion of soluble Pb^{2+} that is left behind in solution becomes significant. The lead isotopes will have their normal proportions, so the mass ratio of $^{208}Pb : {}^{207}Pb$ will be about 2.5 to 1. The isotope dilution technique is to add to the sample a known mass of a lead compound that contains only ^{208}Pb, and precipitate the PbS as before. The greater amount of precipitate makes it easier to collect. Mass spectrometry can now be used to measure the new relative abundances of ^{208}Pb and ^{207}Pb (which will be the same as in the sample they were precipitated from). Knowing the quantity of extra ^{208}Pb which has been added, makes it a simple calculation to work back to the quantity of lead in the ice before enrichment.

Patterson's results are shown in Figure 5.10 (overleaf). They suggest that the lead concentration in snows deposited from the atmosphere increased enormously between 800 BC and AD 1965, but that the increase was especially severe after 1940 — a time when one source of atmospheric contamination, the emission from lead smelters, had been reduced. Patterson's conclusion was simple: the steep increase after 1940 was caused by the huge rise in the consumption of leaded petrol.

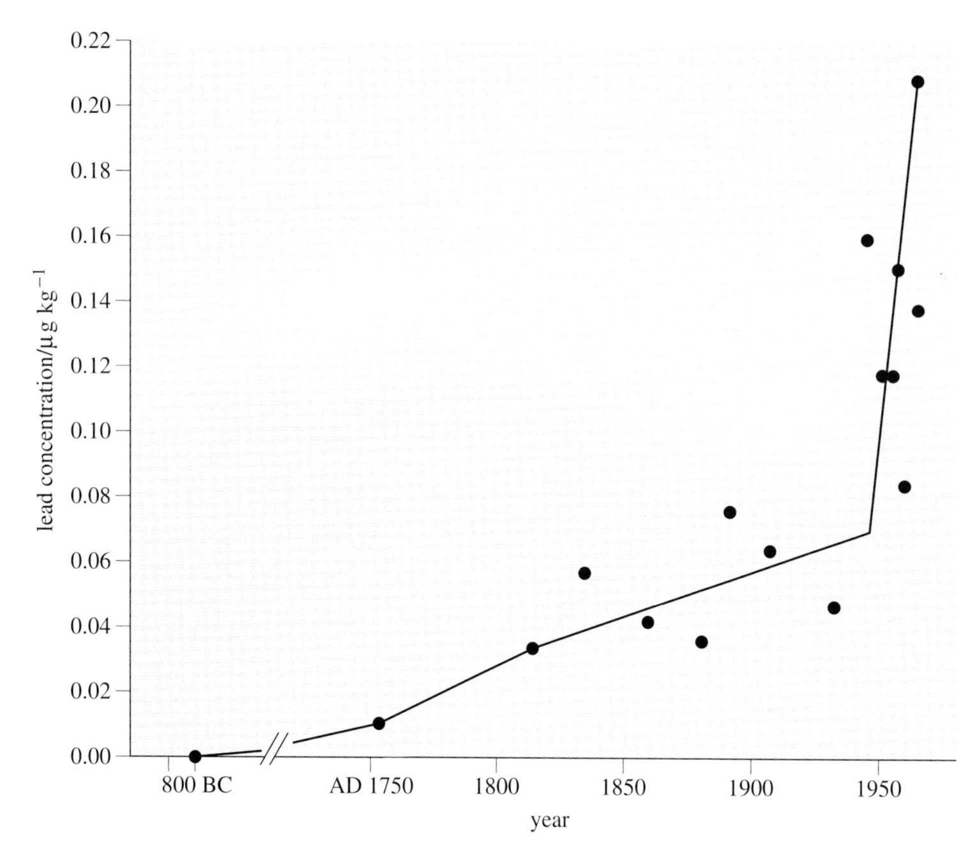

Figure 5.10 The estimated lead concentration of Greenland snows between 800 BC and AD 1965.

When mass spectrometry is used to analyse organic molecules, positively charged ions are created by bombarding the sample with energetic electrons; this tends to remove an electron from the molecule, leaving a positively charged ion known as the **molecular ion**. The bombardment with electrons is of sufficient energy to cause some of the molecules to fragment to smaller charged particles which are called **fragment ions**. As all we have done in producing the molecular ion is to remove one electron, the mass of the original molecule is virtually identical with the mass of the molecular ion, and so we can determine the mass of a molecule, that is, the relative molecular mass. The masses of the fragment ions are also measured in the experiment, and a characteristic pattern is obtained for a particular molecule. The mass spectrum of benzene is shown in Figure 5.11.

There are other types of mass spectrometer besides the ones described here, but they all have the same common feature that they allow species to be sorted and measured according to their mass.

- Do you think that the determination of the relative molecular mass of an unknown substance would yield an unambiguous identification?
- No, there would be far too many possible atomic combinations.

Excluding inspired guesses, we can no more answer the question, 'This molecule has a relative molecular mass of 78; what is it?' than we can answer 'A book weighs

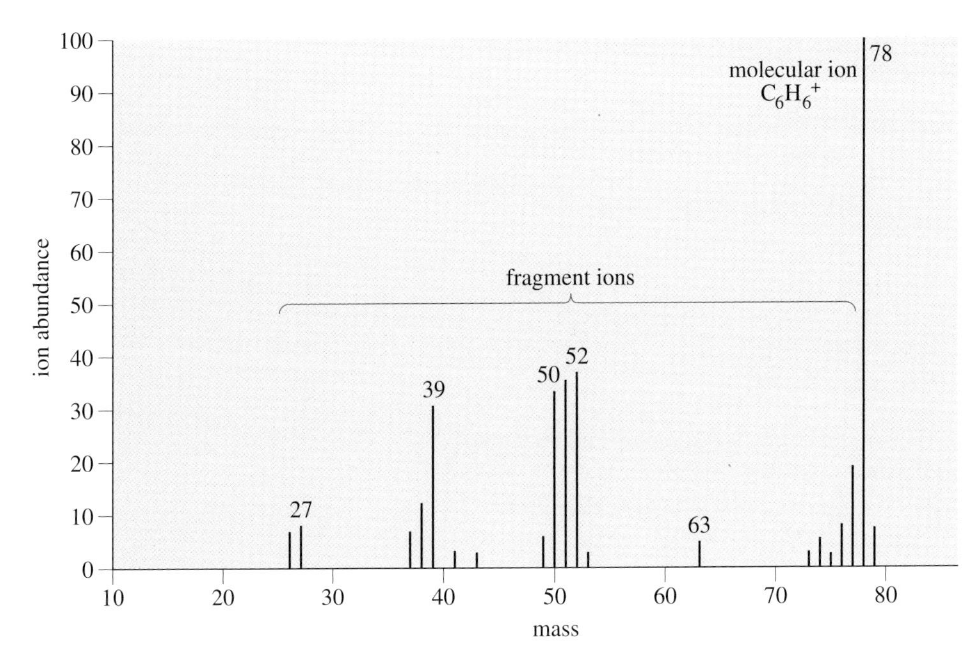

Figure 5.11
The mass spectrum of benzene. Note that the large peak at mass 78 corresponds to benzene's relative molecular mass.

360 g; which book is it?' There are many different molecules with a relative molecular mass of 78, and probably many books that weigh 360 g. But, all is not lost! Different structural features can be identified by the nature of the fragment ions produced. The mass spectrum of benzene, with a relative molecular mass of 78 (Figure 5.11) shows that a mass spectrum is a record of the abundances of the molecular and fragment ions versus their masses. Such a spectrum can be used as a 'molecular fingerprint' to identify a molecule by comparison with a data bank of standard spectra.

COMPUTER ACTIVITY 5.3 Mass Spectrometer

At some point in the near future you should watch the video entitled *Mass Spectrometer* in the multimedia activity *Practical techniques* on the *Experimental techniques* CD-ROM that accompanies this book. This sequence shows you how a mass spectrum is recorded. This activity should take approximately 5 minutes to complete.

But we can use mass spectrometry to do more than this.

- How could you determine if a sample contained 2-methylpropane or 2,3-dimethylbutane?

- This is a trivial task for mass spectrometry. The former weighs in at a relative molecular mass of 58, and the latter at 86. This is a case in which determination of molecular mass to the nearest whole number is perfectly adequate to clear up an ambiguity. Not all tasks are this simple!

- 2-methylpropane has a molecular formula of C_4H_{10}. Is this the only formula with a relative molecular mass of 58 (to the nearest whole number)? See if you can find any others.

- No. Table 5.4 shows some of the molecular formulae with the same integer relative molecular mass.

Table 5.4 Some molecular formulae with relative molecular mass (M_r) of 58, and their accurate masses, calculated by summation of the masses of the most abundant isotopes

Formula	Integer M_r	Accurate M_r
$C_2H_2O_2$	58	58.005 4
CH_2N_2O	58	58.016 7
C_3H_6O	58	58.041 9
$C_2H_6N_2$	58	58.053 2
C_4H_{10}	58	58.078 3

On the relative atomic mass scale, the ^{12}C isotope is given a value of 12.000 0, and the masses of all the other elements are defined relative to this reference. It turns out that isotopes of all of the other elements have relative masses that are slightly greater or slightly less than whole numbers. For example, ^{1}H and ^{16}O have relative masses of 1.007 83 and 15.994 91, respectively. Table 5.4 also gives the accurate relative molecular masses of the formulae shown, calculated by adding together the accurate masses of the principal isotopes of each element present. The general conclusion from Table 5.4 is that molecules with the same integer masses but different molecular formulae have different relative molecular masses when measured to four or five decimal places. Conversely, if we measure accurately the mass of a molecule using a mass spectrometer we can calculate the molecular formula of that molecule.

- Suggest a molecular formula for an unknown compound whose relative molecular mass was measured as 58.041 7.
- Examination of Table 5.4 would suggest C_3H_6O. Notice that there's always a little experimental error in the mass measurement: the calculated and measured values will almost never be exactly the same.

Think back to what we have just been learning about combustion analysis and the information that such analysis gave us — the empirical formula of a compound. We can now combine the results from the two techniques: combustion analysis gives us an empirical formula of C_3H_6O, and from mass spectrometry we know that the compound has a molecular mass of 58.041 7. We can see straight away that the empirical formula must also be the molecular formula, i.e. C_3H_6O.

- Suppose the mass spectrum had given a molecular ion of 116 for this sample?
- Clearly, the empirical formula of C_3H_6O does not correspond to the molecular formula, as this would only give a molecular ion of 58. Instead we need to find what factor by which to scale the empirical formula to give a molecular mass of 116. In this case, multiplying 58 by 2 gives 116. Therefore the empirical formula must also be doubled to give the molecular formula, i.e. the molecular formula is $C_6H_{12}O_2$.

QUESTION 5.5

If a compound has an empirical formula of $C_2H_6O_2N$ and a molecular ion of 228, what is its molecular formula?

The determination of molecular formula in this way is enormously helpful for structure elucidation. For example, an unknown compound with a molecular formula of C_3H_6O cannot possibly be an amine (no nitrogen present), an ester or carboxylic acid (both requiring two oxygen atoms). In other words, the molecular formula, once determined, can be used to eliminate some structural types from consideration.

- What class(es) of compounds is(are) consistent with C_3H_6O?
- Because a single oxygen atom is present, the compound could be an ether, an alcohol, a ketone or an aldehyde.

The next important question to answer is: does the unknown molecule contain any double bonds, triple bonds, or rings? We can answer this question by examining the molecular formula and working out the number of **double-bond equivalents** in the molecule. A saturated acyclic (non-cyclic) hydrocarbon has a molecular formula C_nH_{2n+2}.

A compound with a molecular formula C_nH_{2n} must have one double bond or one saturated ring in it; such a compound is said to contain one double-bond equivalent, for example

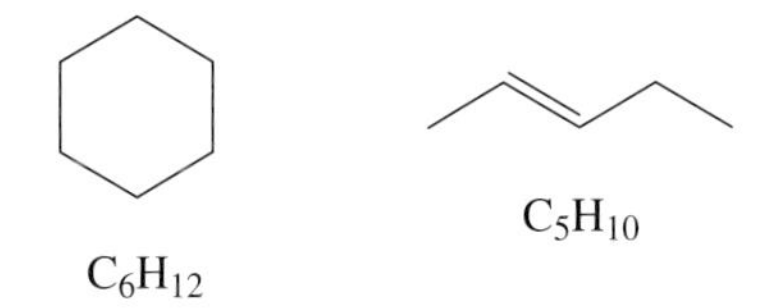

- Check that this is true for the following compounds by working out their molecular formulae

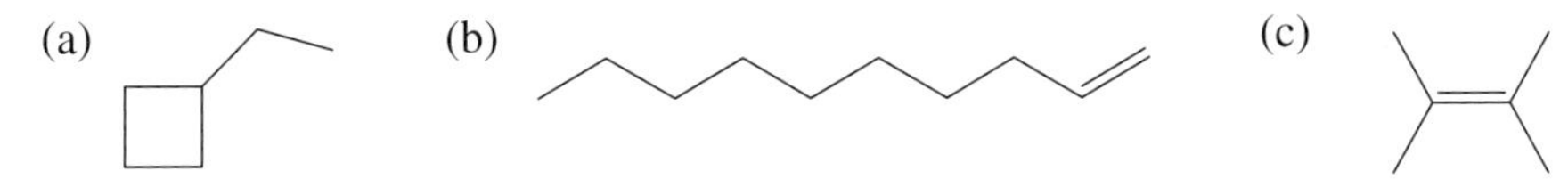

- The molecular formulae are (a) C_6H_{12}; (b) $C_{10}H_{20}$; (c) C_6H_{12}.

A compound of molecular formula C_nH_{2n-2} contains *two* double-bond equivalents; these could be two double bonds, *or* two rings, *or* one ring and one double bond, *or* one triple bond.

- Check that this is so for the following compounds

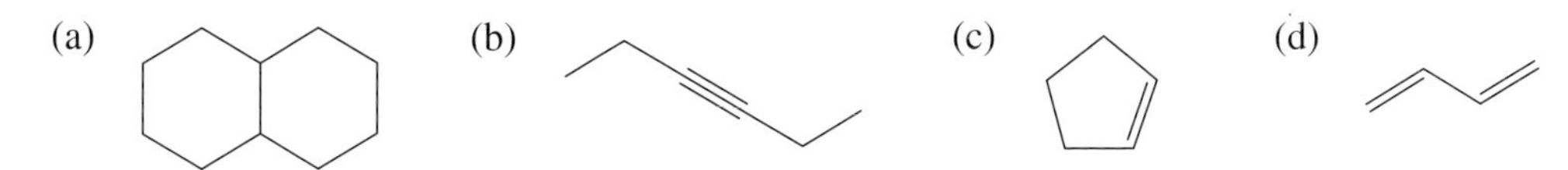

- The molecular formulae are: (a) $C_{10}H_{18}$; (b) C_6H_{10}; (c) C_5H_8; (d) C_4H_6.

  In general, for every two hydrogen atoms less than the fully saturated compound (C_nH_{2n+2}), we have one double-bond equivalent. The presence of an oxygen atom in a molecule does not affect the calculation; all that you need to remember is that oxygen can be part of a double bond (C=O) or part of a ring.

- How many double-bond equivalents has C_3H_6O?

- If we set n equal to 3 and ignore the oxygen, then the saturated compound would be C_3H_8, so the number of double-bond equivalents is one, that is, $(8 - 6)/2$. The molecule must contain one ring, or a C=C bond, or a C=O bond.

Although you can ignore O when working out double-bond equivalents from molecular formulae, you must take N into account. It's not difficult: before working out the number of double-bond equivalents as shown above, assume that each nitrogen in the molecular formula is equivalent to a CH unit (it has the same number of electrons). For example, consider $C_6H_4N_2O_4$. Ignore oxygen altogether and replace each N with CH:

C_6H_4 and 2N is equivalent to

C_6H_4 + 2(CH) which is equivalent to C_8H_6

The fully saturated hydrocarbon with eight carbons would be C_8H_{18}.

Therefore the number of double-bond equivalents in $C_6H_4N_2O_4$ is given by $(18 - 6)/2 = 6$.

QUESTION 5.6

Work out the number of double-bond equivalents in each of the following formulae: (a) $C_6H_{11}N$; (b) $C_{27}H_{46}O$; (c) $C_{10}H_{22}$.

Detailed interpretation of mass spectra is beyond the scope of this book. Also, although taken together, combustion analysis and mass spectrometry can give the molecular mass and empirical and molecular formulae, they do not give us any further information about the compound under study, such as the linkage of the atoms within the molecule. How we solve this will be dealt with in Part 2, *Spectroscopy* (on the CD-ROM).

The other main drawback with both combustion analysis and mass spectrometry is that they are destructive techniques, i.e. we lose our sample; we cannot get it back for use in further tests. Fortunately, only small amounts of sample are required for both tests (about 10^{-9} g for mass spectroscopy and 10 to 20 mg for combustion analysis). So although extremely valuable for the data they deliver, in this respect, they are not ideal techniques.

- List the characteristics that you believe the ideal identification technique should have.
- The ideal technique should be (i) very sensitive, so that only small amounts of substance are required, (ii) non-destructive, (iii) capable of yielding an unambiguous result, (iv) fast, (v) cheap!

- We have already seen that mass spectrometry and combustion analysis do not satisfy all these requirements. Does the determination of the boiling temperature fulfil all these requirements?
- A boiling temperature determination is fairly rapid, it is certainly cheap, it is non-destructive (in most cases), and it can be performed on a fairly small amount of substance. Thus, it fulfils many of the requirements, but fails on one very important respect — the third criterion. There are many compounds with very similar boiling temperatures, and so unambiguous identification is impossible. If we could determine boiling temperatures to within a thousandth

of a degree, then we might be in business; but boiling temperature is very dependent on sample purity and on atmospheric pressure, and these two quantities are very difficult to take into account to this degree of accuracy.

Is there then a technique that we can adopt which fulfils all of these requirements? The answer must be no; life can't be that easy! So identification is normally brought about by the accumulated information from the application of several techniques: the more data we collect, the greater the chance of unambiguously fixing the identity of the substance. The degree of sophistication required, however, depends on the degree of uncertainty in the molecular identity. If we were fairly confident about molecular identity, then we might try to confirm our suspicions by the determination of the boiling temperature; however, the wise chemist would not stop there, but would seek confirmation by the application of some further technique.

The technique of *X-ray crystallography* * has been employed with increasing frequency in recent years. Modern X-ray diffractometers and computational methods can determine the crystalline structure of a compound relatively routinely, often in a day or less, whereas not many years ago, a structure determination could take weeks or even months. This technique gives a fully determined structure for a compound, with each atomic position, and therefore all the bond lengths and angles, determined. Such instruments are, however, not cheap and not available everywhere. It also frequently quite difficult to make a suitable single crystal, so the technique cannot always be used. However, when it is available and possible, it is the technique of choice.

There is one set of techniques that is very sensitive, is fast, goes a long way to giving an unambiguous identification, is non-destructive *and* is widely available. This set of techniques is called **spectroscopy**. Spectroscopy provides one of the most versatile and widely used methods of molecular identification. Although the instrumentation associated with some forms of spectroscopy can be far from cheap — anything up to £500 000 per instrument at 2002 prices — the cost per substance identified is very low. One instrument can record many thousands of spectra in its lifetime, and, since identification via this process is made relatively quickly, there is a huge saving in time, and saving time saves money. The remainder of the teaching in this book now continues in Part 2 on the *Spectroscopy* CD-ROM.

5.4 Summary of Section 5

1 There are many methods available to help with identifying a compound.

2 Elemental analysis is an excellent technique for identifying the elements present in a compound. Carbon, nitrogen, and hydrogen are determined by combustion analysis; sulfur is determined by burning in oxygen and reaction with hydrogen peroxide. Nitrogen can also be determined by the Kjeldahl method. The oxygen content of a compound is usually determined by difference.

3 Elemental analysis is a destructive technique.

4 Spectroscopy offers an alternative non-destructive technique, and takes many forms. Every element has a unique atomic absorption and emission spectrum.

5 Elemental analysis and spectroscopy may be used together to determine the empirical formula of a compound.

* X-ray crystallography is discussed briefly in *The Third Dimension*[5].

6 Mass spectrometry can help determine the molecular mass and thus the molecular formula.

7 An ideal identification technique should be sensitive, non-destructive, unambiguous, fast and cheap. Unfortunately there is no ideal technique but chemists instead use a combination of methods in order to determine the structure of a compound.

QUESTION 5.7

On reacting compound A with an excess of aqueous sodium hydroxide, a compound B is obtained. Combustion analysis of B gave C 53.31% and H 11.18%. Mass spectrometry gave a molecular ion of value 90.12 and fragment peaks at mass numbers 73.11 and 56.10.

Suggest

(a) the empirical formula of B,

(b) the molecular formula of B,

(c) a possible structure for B,

(d) a possible structure for starting material A.

QUESTION 5.8

Calculate the empirical formula of a compound containing C 62.07% and H 10.34%.

QUESTION 5.9

A molybdenum complex contains one molybdenum (18.6%) and two iodines (49.2%), as well as carbonyl (CO) and acetonitrile (CH_3CN) ligands. The combustion analysis gives C 16.3%, H 1.2% and N 5.4%. Determine the empirical formula. Assuming that the molecular formula is the same as the empirical formula, determine the number of acetonitrile and carbonyl ligands.

QUESTION 5.10

When $[WI_2(CO)_3(NCCH_3)_2]$ is reacted with triphenyl stibine, $Sb(C_6H_5)_3$, only one of the acetonitrile ligands is replaced by the stibine. Determine the percentages of C, H, N and O in the new complex.

6

CONCLUSION TO PART 1

You should now be able to plan a basic chemical reaction, to know the types of apparatus that you would need to employ and have an idea about how to perform the reaction. Hopefully you will also now be aware of the vast armoury of tools the modern chemist has available for the purification and identification of compounds generated during a chemical reaction. It is worthwhile at this point thinking back to a point that we made in the introduction to Part 1. Much of the chemistry that you have learnt thus far has been known for many years and was discovered by the early, pioneering chemists. While apparatus and glassware may not have changed much over the years, these early chemists had none of the modern methods of compound identification and characterization available to them. It is to their considerable credit that they made the compounds and scientific advances that they did, given the extremely difficult circumstances under which they were working. Modern chemists have far more techniques available, but against that, the compounds that the modern chemist typically analyses tend to be far more complex than the early chemists would have ever considered tackling.

In Part 2, you will now use the *Spectroscopy* CD-ROM to tackle the principles of vibrational (infrared and Raman) and NMR spectroscopies and explain how they may be used in molecular identification.

LEARNING OUTCOMES FOR PART 1

When you have completed Part 1, you should be able to:

1 Define in your own words, recognize valid definitions of, and use in a correct context, the scientific terms, concepts and principles listed in the following table. (All questions)

List of scientific terms, concepts and principles used in Part 1 of this book.

Term	Page number	Term	Page number
absorbance	61	mass spectrum	66
absorption band	CD	molecular formula	63
adsorbent	41	molecular ion	68
adsorption	41	polar molecule	24
anhydrous	26	purity	54
atomic spectroscopy	58	Quickfit apparatus	12
chromatography	41	recrystallization	50
column chromatography	41	reflux	19
combustion analysis	56	retardation factor R_f	41
COSHH	16	retention time	47
distillation	34	risk assessment	16
double-bond equivalent	71	separation	23
eluant	41	solubility	23
elution	41	solvent extraction	23
fraction	45	spectroscopy	73
fractional distillation	35	thin-layer chromatography (TLC)	41
fragment ion	68	wavenumber	CD
gas-liquid chromatography (GLC)	46	work-up	21
high-performance liquid chromatography (HPLC)	46		

2 Be familiar with the language of chemistry and the way in which scientists report and write-up their experiments. (Question 1.1)

3 Be familiar with common laboratory glassware and apparatus. Be able to write an experimental report in an appropriate style, and be able to follow a written experimental procedure. (Questions 2.1 to 2.4)

4 Describe the common techniques of separation and purification, and indicate the most appropriate technique to use for the separation of a given mixture. (Questions 2.1 to 2.10, 4.1)

5 Compare and contrast the common methods of separation and purification. (Questions 2.2 to 2.10, 4.1)

6 List the features of an ideal method for the identification of molecules via comparison with standard data, and describe how a given physical measurement compares with this ideal. (Question 4.1)

7 Interconvert given values of energy, frequency, wavelength and wavenumber units. (Question 5.1)

8 Be able to calculate reaction yields, empirical and molecular formula, double-bond equivalents and elemental percentages from numerical data. (Questions 5.2 to 5.10)

QUESTIONS: ANSWERS AND COMMENTS

QUESTION 1.1 (*Learning Outcome 2*)

There is obviously a lot wrong with this particular experimental write-up!

First, the student should have identified what the three chemicals were, stated the amounts of each one used and if there were any particular hazards or safety issues connected with the reagents. The flask was a round-bottomed flask and the white plastic bar was an internal magnetic stirrer. The student should also have stated from the outset that the reaction had to be performed under dry conditions and what precautions were taken to ensure that it stayed dry. What temperature was the reaction performed at? We have no indication of what happened after 35 minutes and if it was gradual or sudden. Similarly, when the student stopped the reaction, how was this done? Were other reagents added? What were the solvents the student added? Which layer contained the product? And what does 'lots of' equate to in terms of mass or percentage yield?

If we call our three starting materials X, Y and Z, then a better experimental procedure might read something like:

X (number of g, number of moles), Y (number of g, number of moles) and Z (number of g, number of moles) were placed in a clean, dry, nitrogen-flushed round-bottomed flask, equipped with a magnetic stirring bar. Nitrogen was bubbled slowly through the apparatus throughout the experiment. The reaction was stirred at room temperature for 35 minutes, during which time the solution changed from clear to a pale yellow colour. Water (volume) was carefully added to the reaction, followed by ethoxyethane (diethyl ether) (volume). The solution was transferred to a separating funnel and the lower aqueous layer removed. The upper organic layer was washed with water (volume) and then dried using anhydrous magnesium sulfate. The filtered organic layer was evaporated to yield the crude product as a yellow oil (amount, g or moles, percentage yield). The product was purified by … (state purification method(s) used).

QUESTION 2.1 (*Learning Outcomes 3 and 4*)

Water and ethoxyethane are immiscible and form two layers in the separating funnel; the ethoxyethane has a lower density than water and so forms the top layer, with the aqueous layer at the bottom.

Remember that the key concept of solvent separation is that like dissolves like. Therefore any organic compounds tend to dissolve in the ether, and ionic compounds dissolve in the water. From the reaction mixture, toluene, and benzyl bromide dissolve in the ethoxyethane, which is the top layer in the separating funnel, whereas sodium chloride, potassium bromide and sodium hydrogen carbonate are found in the aqueous (bottom) layer.

QUESTION 2.2 (*Learning Outcomes 3, 4 and 5*)

Reflux a mixture of pentane-1,5-dinitrile (insert the number of g and mol to be used) and sulfuric acid (ml) in water (ml) for 10 hours and allow to cool. Using a separating funnel, extract the solution with ethoxyethane (4×150 ml portions). Dry the ethereal extracts with anhydrous magnesium sulfate. Distil off the ether to yield the desired product, pentane-1,5-dioic acid.

QUESTION 2.3 *(Learning Outcomes 3, 4 and 5)*

The procedures adopted in this preparation are very similar to those used in the preparation of 2-chloro-2-methylpropane discussed in Section 2.1. At the end of the reaction there are two liquid layers present — an organic layer containing the dibromide and any other water-insoluble substances such as starting material, and an aqueous layer containing sulfuric acid, any unchanged HBr and other water-soluble substances. A separation of organic from inorganic components is possible without the addition of solvents. By separating the layers using a separating funnel, we separate the required product from the inorganic material. By adding water and 'washing' the organic layer, we effectively improve the organic/inorganic separation. The 1,5-dibromopentane is then separated from other organic material by distillation (see Section 2.2) after removing traces of water with the drying agent, anhydrous magnesium sulfate.

QUESTION 2.4 *(Learning Outcomes 3, 4 and 5)*

At the start of the reaction there will be two liquid phases present: an organic layer of pentanenitrile, and an inorganic/aqueous layer containing sodium hydroxide and water. The product of the reaction is an acid but, because this is being formed in an alkaline (NaOH) solution, the product will be the sodium salt of the acid. This salt will be soluble in water. Therefore the reaction is continued until all the organic layer disappears; at this point, all the pentanenitrile will have been consumed. At the end of the reaction there will be one aqueous layer containing the required product (as the sodium salt) plus all the other possible products and unused reagents, etc. How then is this organic acid separated from the aqueous layer? The aqueous solution is acidified with sulfuric acid; this produces the organic acid, which is insoluble in water and forms a separate liquid layer. The solution is transferred to a separating funnel, and the aqueous layer is run off and set aside. The upper layer — almost pure pentanoic acid — is dried with anhydrous calcium sulfate, and the remaining solution is distilled to remove the final traces of impurity.

However, often after the addition of sulfuric acid, the organic acid does not form a separate layer but remains as a suspension in the water. In this case, the organic acid is usually removed from the inorganic layer by shaking with an organic solvent such as ethoxyethane; that is, a simple solvent extraction is used.

QUESTION 2.5 *(Learning Outcomes 4 and 5)*

An R_f value outside the range 0.2 to 0.8 may not indicate the true chromatographic behaviour of the substance. Compounds running close to the solvent front may never have been properly adsorbed onto the coating of the plate, and compounds staying near to the baseline may have moved only under the influence of other components of the mixture or unevaporated traces of the solvent used to dissolve the mixture. Also, outside this range, mixtures of fast-running or slow-running compounds may remain unresolved.

QUESTION 2.6 *(Learning Outcomes 4 and 5)*

The amount of material applied is not critical, although small spots tend to give better results, and this makes TLC a very good technique to use.

QUESTION 2.7 *(Learning Outcomes 4 and 5)*

(i) TLC is routinely used in analytical laboratories to detect and identify components of mixtures, for example drugs in a forensic or hospital laboratory, or food additives (flavourings and colourings) in the Laboratory of the Public Analyst.

(ii) A natural products chemist could use thin-layer chromatography to discover how many components a plant extract contained, and to compare extracts from different plants of the same family, or extracts from, say, the roots, leaves and flowers of a single plant species.

(iii) The synthetic chemist uses TLC for quality control of the products isolated at the end of a reaction. The technique can also be used while the reaction is running, to check the extent to which starting materials have been used up, or to establish whether the reaction is complete.

QUESTION 2.8 *(Learning Outcomes 4 and 5)*

Compare your completed table with Table Q.1. Don't be too alarmed if your version does not correspond exactly with ours, because the questions are fairly subjective.

Table Q.1 Completed table for choice of separation techniques

	Distillation	Solvent Extraction	Chromatography
1 Will separate a small number of components in a mixture	✔✔	✔✔	✔✔✔
2 Will separate a large number of components	✔	✔	✔✔✔
3 Will separate small amounts of material	✔✔	✔✔	✔✔✔
4 Will separate large amounts of material (i.e. on an industrial scale)	✔✔✔	✔✔✔	✔

The important points to note are that for the separation of small amounts of a complex mixture, chromatography is supreme. The technique is simple to operate, comparatively cheap to run, and is quick, especially in the case of gas-liquid chromatography (GLC). As the amount of material increases, or the complexity of the mixture decreases, the other techniques increase in applicability and become most powerful as methods for the industrial separation of simple mixtures (sum the ticks in each column in rows 1 and 4). Distillation and solvent extraction at the industrial level are fairly comparable in speed, cost and simplicity.

QUESTION 2.9 *(Learning Outcomes 4 and 5)*

If we wish to purify an organic liquid, the implication is that only a few components are present. Therefore we require a technique to separate a small number of components in large amounts. This can be found from Table Q.1 by looking for the technique that gives the greatest number of ticks for rows 1 and 4 combined. This can be either distillation or solvent extraction. Further information would be required to select the more appropriate of these two techniques. If the main impurity is water, then drying over an anhydrous inorganic solid might be most appropriate.

QUESTION 2.10 *(Learning Outcomes 4 and 5)*

There are only two components present, benzene and cyclohexane. To separate small amounts of this mixture we sum the ticks in rows 1 and 3 of Table Q.1 to reveal the most appropriate technique: chromatography beats distillation and solvent extraction. We have some additional information in the question here; the boiling temperatures of the two components are very close, and so separation could not be effected even by fractional distillation (which needs at least 2 °C difference in boiling temperatures). This confirms chromatography as the preferred technique, although even this may not be easy, as both solvents have similar polarity. As both components are relatively volatile, gas-liquid chromatography would be the first choice.

QUESTION 4.1 *(Learning Outcomes 4, 5 and 6)*

If we wish to check the purity of a sample, we might favour a technique that will reveal the number of compounds present, and give some guide as to their relative amounts. If we are checking for purity we must assume that the substance is relatively pure, and therefore we are looking for a technique that will separate small amounts of a mixture with a relatively small number of components. This can be found from Table Q.1 by summing the ticks in rows 1 and 3: chromatography is the choice.

(i) For an organic solid, high-performance liquid chromatography is probably the ideal technique, although thin-layer chromatography would be favoured for being quick, simple and cheap. If the solid is relatively volatile, then gas-liquid chromatography (GLC) could be used.

(ii) For an organic liquid, GLC is probably the ideal technique. The liquid at room temperature would be injected into a column at elevated temperatures, and would be carried through the column by a gaseous 'solvent' such as nitrogen gas. High-performance liquid chromatography, HPLC, may also provide the required information.

Alternatively, the boiling temperature of a liquid, or the melting temperature of a solid, might be measured. The temperatures will be sharp and agree with published values only if the compounds are pure. A common purity check involves combustion analysis (see Section 5.1). If the experimentally determined empirical formula matches that of the required organic compound, it is likely to be pure. With all three of these methods, if the temperature or formula turns out to be incorrect, the sample is impure. However, there is no way of identifying the impurities using these methods.

QUESTION 5.1 *(Learning Outcome 7)*

(i) Converting from eV to J: $\Delta E = 2.1 \times 1.602 \times 10^{-19}\,\text{J} = 3.364 \times 10^{-19}\,\text{J}$.

Planck's constant, $h = 6.626 \times 10^{-34}\,\text{J s}$, (to four sig. figs.), so using $\Delta E = h\nu$

$$3.364 \times 10^{-19}\,\text{J} = (6.626 \times 10^{-34}\,\text{J s}) \times \nu$$

therefore $\nu = 3.364 \times 10^{-19}\,\text{J}/6.626 \times 10^{-34}\,\text{J s}$

$$= 5.077 \times 10^{14}\,\text{s}^{-1}, \text{ or } 5.1 \times 10^{14}\,\text{s}^{-1} \text{ to 2 sig figs.}$$

(ii) Converting $5.1 \times 10^{14}\,\text{s}^{-1}$ to wavelength we use $c = \nu\lambda$ where c, the velocity of light, is $3.0 \times 10^{8}\,\text{m s}^{-1}$.

$$\lambda = 3.0 \times 10^{8}\,\text{m s}^{-1}/5.1 \times 10^{14}\,\text{s}^{-1} = 5.9 \times 10^{-7}\,\text{m} = 590\,\text{nm}.$$

The *emitted* orange light is thus due to the $3p \longrightarrow 3s$ transition (Figure 5.4b).

QUESTION 5.2 *(Learning Outcome 8)*

44 g of carbon dioxide contain 12 g of carbon

$$\therefore 0.0792\,\text{g of } CO_2 \text{ contain } \frac{12}{44} \times 0.0792 = 0.0216\,\text{g of carbon}$$

$$\therefore \%\text{ carbon in compound} = \frac{0.0216}{0.0234} \times 100 = 92.3\%$$

18 g of water contain 2 g of hydrogen

$\therefore$ 0.0162 g of water contain $\frac{2}{18} \times 0.0162 = 0.0018$ g of hydrogen

$\therefore$ % hydrogen in compound $= \frac{0.0018}{0.0234} \times 100 = 7.7\%$

The oxygen content of a compound is not usually detected by elemental analysis, and is normally determined by difference, i.e. the remainder when all the other percentages have been subtracted from 100. In this case the percentages of carbon and hydrogen total 100%, so there is no oxygen present, and we are dealing with a hydrocarbon.

QUESTION 5.3 *(Learning Outcome 8)*

This time, the percentages add up to 100% and so there is no oxygen present. Once again we tabulate our calculation.

Table Q.2 For Question 5.3

Element	C	H
Percentage by mass	81.8	18.2
Number of moles	$\frac{81.8}{12}$	$\frac{18.2}{1}$
Atom ratios	6.82	18.2
Dividing through by the lowest atomic ratio	$\frac{6.82}{6.82}$	$\frac{18.2}{6.82}$
Whole number ratio of atoms	1	2.67

Since the carbon to hydrogen ratio of 1 : 2.67 is too far from a whole number to be attributed to experimental error, the empirical formula must be the lowest multiple to give a whole number ratio, and 2 × (1 : 2.67) gives 2 : 5.34, still not a whole number, but 3 × (1 : 2.67) gives 3 : 8.01, which is a whole number ratio within the experimental error. It is difficult to be precise about the experimental error that is acceptable in determinations of this kind, but the percentage of C, H, or N determined by combustion analysis is usually quoted to 0.1%.

Therefore the empirical formula is C_3H_8.

QUESTION 5.4 *(Learning Outcome 8)*

^{208}Pb is most abundant and ^{204}Pb is the least abundant — they have the largest and smallest peaks respectively. The peak heights for ^{208}Pb and ^{207}Pb are in the ratio 2.5 to 1, so there are five atoms of ^{208}Pb for every two atoms of ^{207}Pb.

QUESTION 5.5 *(Learning Outcome 8)*

The empirical formula corresponds to a mass of 76, clearly short of the desired molecular ion. Dividing 228 by 76 gives us a factor of three for scaling the empirical formula. Therefore the molecular formula of this compound is $(3 \times C_2H_6O_2N) = C_6H_{18}O_6N_3$.

QUESTION 5.6 *(Learning Outcome 8)*

(a) After we substitute CH for N, $C_6H_{11}N$ becomes C_7H_{12}. The saturated C_7 hydrocarbon would be C_7H_{16}.

∴ Number of double-bond equivalents = (16 – 12)/2 = 2.

(b) After ignoring O, we are left with $C_{27}H_{46}$. The saturated C_{27} hydrocarbon would be $C_{27}H_{56}$.

∴ Number of double-bond equivalents = (56 – 46)/2 = 5.

This is the molecular formula of cholesterol, which occurs in cream and human arteries amongst other places. It has one double bond and four rings, so its formula should reveal five double-bond equivalents.

(c) You should have spotted that $C_{10}H_{22}$ is the formula of a saturated, non-cyclic alkane C_nH_{2n+2}, where n = 10. Therefore, without bothering with maths, we can say that this molecule has no double-bond equivalents.

QUESTION 5.7 *(Learning Outcome 8)*

(a) The empirical formula of B is derived in Table Q.3. By difference, the oxygen content must be (100 – (53.31 + 11.18)) = 35.51% oxygen.

Table Q.3 For Question 5.7

Element	C	H	O
Percentage by mass	53.31	11.18	35.51
Number of moles	$\frac{53.31}{12}$	$\frac{11.18}{1}$	$\frac{35.51}{16}$
Atom ratios	4.44	11.18	2.22
Dividing through by the lowest atomic ratio	$\frac{4.44}{2.22}$	$\frac{11.18}{2.22}$	$\frac{2.22}{2.22}$
Whole number ratio of atoms	2	5.03	1

(a) Allowing for experimental error, the empirical formula of compound B is C_2H_5O.

(b) The empirical formula of B is C_2H_5O. If we sum the atomic masses, this gives a relative molecular mass for B of 45. However, we know from mass spectrometry that the molecular ion of B occurs at 90.12. Therefore the empirical formula cannot be the molecular formula for B. Instead, as we can see, it is one-half the relative molecular mass of B. Therefore the molecular formula must be double the empirical formula, i.e. $C_4H_{10}O_2$.

(c) B contains no double-bond equivalents. There are many possible compounds that could be drawn for B incorporating two oxygen-containing functional groups. Notice however that the mass spectrum shows two fragment peaks, both corresponding to losses of 17 mass units. Both of these may be attributed to loss of an alcohol group, —OH, which has a mass of 17. Therefore we are probably looking for a molecule containing two alcohol groups such as

$C_4H_{10}O_2$
R.M.M. = 90.12
C, 53.31; H, 11.18; O, 35.51

(d) The reaction conditions described correspond to an S_N2 reaction. You should recall that primary alkyl halides are by far the best substrates for this type of reaction. Therefore the most likely substrate would be a primary alkyl halide. However, as we are producing two alcohols in the product, the starting material must contain two primary alkyl halides. Therefore a possible structure for A is

Br Br

QUESTION 5.8 *(Learning Outcome 8)*

By difference, the oxygen content must be (100 – (62.07 + 10.34)) = 27.59% oxygen.

Table Q.4 For Question 5.8

Element	C	H	O
Percentage by mass	62.07	10.34	27.59
Number of moles	$\frac{62.07}{12}$	$\frac{10.34}{1}$	$\frac{27.59}{16}$
Atom ratios	5.17	10.34	1.72
Dividing through by the lowest atomic ratio	$\frac{5.17}{1.72}$	$\frac{10.34}{1.72}$	$\frac{1.72}{1.72}$
Whole number ratio of atoms	3.01	6.01	1.00

Allowing for experimental error, the empirical formula of the compound is C_3H_6O.

QUESTION 5.9 *(Learning Outcome 8)*

The C, H, N and O must account for (100 – (18.6 + 49.2)) = 32.2%. Within this allowance, O accounts for (32.2 – (16.3 + 1.2 + 5.4)) = 9.3%.

Table Q.5 For Question 5.9

Element	C	H	N	O	Mo	I
Percentage by mass	16.3	1.2	5.4	9.3	18.6	49.2
Number of moles	$\frac{16.3}{12}$	$\frac{1.2}{1}$	$\frac{5.4}{12}$	$\frac{9.3}{16}$	$\frac{18.6}{96}$	$\frac{49.2}{127}$
Atomic ratios	1.36	1.20	0.39	0.58	0.194	0.387
Dividing through by the lowest atomic ratio	$\frac{1.36}{0.194}$	$\frac{1.20}{0.194}$	$\frac{0.39}{0.194}$	$\frac{0.58}{0.194}$	$\frac{0.194}{0.194}$	$\frac{0.387}{0.194}$
Whole number ratio of atoms	7	6	2	3	1	2

The empirical formula is $C_7H_6N_2O_3I_2Mo$. Further spectroscopic evidence is necessary to show that the molecule is a seven-coordinate complex: $[MoI_2(CO)_3(NCCH_3)_2]$.

QUESTION 5.10 *(Learning Outcome 8)*

The molecular formula of the new complex is $[WI_2(CO)_3(NCCH_3)Sb(C_6H_5)_3]$. The empirical formula is therefore: $C_{23}H_{18}NO_3I_2SbW$.

We need to determine the relative molecular mass of the complex, and the contribution of each element (working to one decimal place). This is tabulated below, with the percentage contribution calculated in the final column.

Table Q.6 For Question 5.10

Number of atoms of each element	Relative atomic mass	Total mass due to element	%
23 C	12.0	276	$\frac{276}{915.5} \times 100 = 30.2$
18 H	1.0	18	$\frac{18}{915.5} \times 100 = 2.0$
1 N	14.0	14	$\frac{14}{915.5} \times 100 = 1.5$
3 O	16.0	48	$\frac{48}{915.5} \times 100 = 5.2$
2 I	126.9	253.8	$\frac{253.8}{915.5} \times 100 = 27.7$
1 Sb	121.8	121.8	$\frac{121.8}{915.5} \times 100 = 13.3$
1 W	183.9	183.9	$\frac{183.9}{915.5} \times 100 = 20.1$
Relative molecular mass of $C_{23}H_{18}NO_3I_2SbW$		915.5	100

FURTHER READING

1 SXR205, *Exploring the Molecular World*, The Open University (2002).

2 M. Mortimer and P. G. Taylor (eds), *Chemical Kinetics and Mechanism*, The Open University and the Royal Society of Chemistry (2002).

3 P. G. Taylor (ed), *Mechanism and Synthesis*, The Open University and the Royal Society of Chemistry (2002).

4 J. M. Gagan and P. G. Taylor (eds), *Alkenes and Aromatics*, The Open University and the Royal Society of Chemistry (2002).

5 J. M. Gagan and L. E. Smart (eds), *The Third Dimension*, The Open University and the Royal Society of Chemistry (2002).

ACKNOWLEDGEMENTS

Grateful acknowledgement is made to the following sources for permission to reproduce material in this book:

Figure 2.7a: Courtesy of Speyside Distillery; *Figure 2.7b*: Courtesy of Bushmills; *Figure 2.8*: © BP plc (2002); *Figure 2.15*: James Holme/Cellmark Diagnostics/ Science Photo Library; *Figure 5.9*: Courtesy of CalTech.

Every effort has been made to trace all the copyright owners, but if any has been inadvertently overlooked, the publishers will be pleased to make the necessary arrangements at the first opportunity.

Part 2

Spectroscopy

Lesley Smart and Eleanor Crabb

PREAMBLE

The material for studying *Separation, Purification and Identification, Part 2: Spectroscopy* is *all* contained on the CD-ROM. There are eleven multimedia activities available together with their summaries and learning outcomes, and a set of questions.

This suite of programs explores the theory and practice of three spectroscopic techniques; infrared spectroscopy, Raman spectroscopy and nuclear magnetic resonance (NMR) spectroscopy. These techniques are used frequently by chemists to give an insight into the shapes and bonding of molecules. We suggest you work through the programs in the following order:

1. *Introduction to spectroscopy* starts by looking at the properties of electromagnetic radiation. Emission and absorption processes in atomic spectroscopy are revised. The interaction of radiation with a vibrating diatomic molecule is considered for both infrared and Raman spectroscopy.
2. *Practical infrared spectroscopy* is a video sequence showing the operation of a modern infrared spectrometer.
3. *The harmonic oscillator model* allows you to explore the factors affecting molecular vibration frequencies using both a computer-based simulation and a video of a real experiment.
4. *Theory of vibrational spectroscopy* considers the energy of a vibrating molecule and the selection rules governing absorption and emission processes. We calculate the number of normal modes of vibration for linear and non-linear (bent) molecules, and view a computer simulation of the normal modes for both linear and bent triatomic molecules.
5. *Diatomic vibrations* is intended to be used for the comparison of the vibrational behaviour of some common bonds as they vary with temperature.
6. *Interpreting infrared spectra* introduces you to the process of interpreting the infrared spectra of organic molecules by assigning peaks in the spectra to particular groups in the molecule.
7. *Inorganic spectroscopy examples* applies the spectroscopic techniques studied to inorganic examples.
8. *Interpreting NMR spectra* briefly covers the basic theory of nuclear magnetic resonance (NMR) spectroscopy and the process of interpreting the ^{13}C NMR spectra of simple organic compounds.
9. *Practical NMR spectroscopy* is a video sequence showing the operation of a modern NMR spectrometer.
10. *The Java Molecular Editor* (*JME*) is used to input answers to questions in the multimedia activities. *Java* is copyright of Novartis AG, Switzerland.
11. *Integrated spectroscopy* brings together skills learned in the other spectroscopy activities with a set of spectroscopy problems involving more than one technique.

Questions is a set of interactive self-assessment questions which you may use to test your understanding of *Separation, Purification and Identification Part 2*.

Summaries and learning outcomes summarize the material in Part 2, and are provided in a form that can be printed off. They are intended as a revision aid.

Use the tabs on the sides of the computer screen to select which activity you wish to run. Where there are bars on the tabs, these will fill with red as you complete the programs, to give you an 'at a glance' view of your progress.

That completes the *Preamble*. When you have completely studied all the material on the CD-ROM, return to this text for the case study on *Forensic Science*.

CONTENTS OF THE SPECTROSCOPY CD-ROM

Case Study

Forensic Science

Andy Platt, Anya Hunt
and Lesley Smart

INTRODUCTION

1

In its broadest definition, forensic science is the application of science to law. Many of the physical and spectroscopic techniques used in forensic science are covered in the earlier sections of this book. The series of cases described below will help to show how these techniques are used in forensic investigations.

A large number of individuals have been cited as contributing to the early development of forensic science. In 1814 Mathieu Orfila published the first known treatise on the detection of poisons and their effects on various animals. Later, in 1879, Alphonse Bertillon began to develop the science of anthropometry, a systematic procedure of taking a series of body measurements as a means of distinguishing individuals. This was used for about 20 years before being abandoned in favour of Francis Galton's fingerprint technology. Galton's results, published in 1892, were the first statistical proof supporting the uniqueness of fingerprints as personal identification, and fingerprints were first used in a trial in the UK in 1902. Several million sets of fingerprints have now been catalogued and no two sets have been found to be identical· It was, however, Hans Gross who first published (1892) a detailed description of how the application of scientific disciplines could be used successfully in the field of criminal investigation. His work detailed the usefulness of microscopy, chemistry, physics, mineralogy, anthropometry and fingerprinting as tools for the criminal investigator.

In the early 1900s Edmond Locard began to develop forensic science as it is known today. In 1910 he persuaded the Lyon Police Department to give him rooms and assistance to start the first police laboratory. It was Locard's belief that when a criminal came into contact with an object or person that a cross-transfer of evidence occurred - this is known as *Locard's exchange principle*. Thus every criminal could be linked to a crime scene by particles transferred.

The function of the forensic scientist today is largely based around Locard's exchange principle. The expertise available in an operational forensic science laboratory covers a range of disciplines and uses a number of scientific techniques. Forensic scientists must therefore be skilled in many scientific areas. They must also be aware of the demands and constraints of the legal system, so that the results of analysis satisfy the criteria of admissibility as evidence that have been established by the courts.

CASE 1: THE ATLANTA CHILD MURDERS

2

2.1 The incidents

Between 1979 and 1981, 28 young black people were slain in Atlanta, Georgia. On 26 February 1982, a Fulton County Superior Court jury returned a verdict of 'guilty as charged' on two counts of murder brought against Wayne Bertram Williams. Williams had been on trial for the murder by asphyxiation, in April and May 1981, of two young blacks, Nathaniel Cater and Jimmy Ray Payne. During the trial, evidence linking Williams to those murders and to the murders of ten other boys or young men was produced, and immediately after Williams' conviction the district attorney closed 21 of the other cases.

2.2 The investigation

The 'Atlanta child murders' were linked by the presence of some yellow-green nylon fibres and some purple acetate fibres on the body or clothing of the victims. In February 1981 an Atlanta newspaper published an article that reported that several different fibres had been found on a number of the murder victims. Following the publication of this article further bodies recovered from Atlanta rivers were either nude or dressed only in underpants. This was a blow to the investigation since it drastically reduced the likelihood of fibres being found adhering to a body.

An essential part of the case against Williams was the association of the fibres removed from the bodies of the twelve murder victims with similar ones from his home. Fibre evidence is often used to corroborate other evidence in a case, but this was not the situation in the Williams' case, rather the reverse: there were no eye-witnesses to any of the murders, and although other evidence and aspects of the trial were important, they were used to support and complement the fibre evidence.

The yellow-green fibres were very coarse, and microscopy revealed that they had an unusual delta-wing cross-sectional appearance with two long lobes and one short one (Figure 2.1). The thickness indicated that they may have originated from a rug or carpet and that tracing their source could be possible.

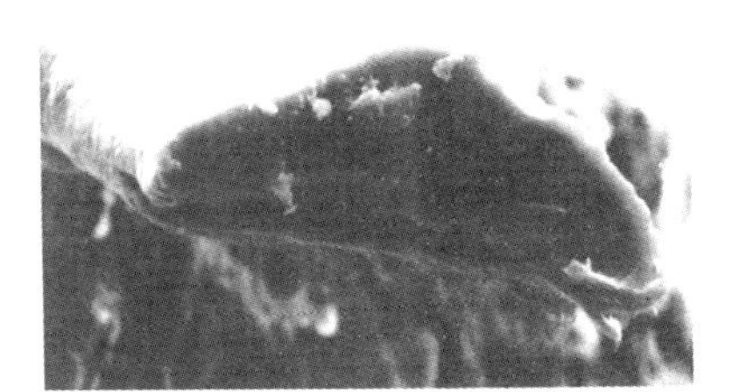

Figure 2.1
A scanning electron micrograph of the cross-section of a fibre removed from a sheet used to transport the body of a murder victim.

On 24 May 1981, the nude body of Nathaniel Cater was pulled from the Chattahoochee River approximately one mile downstream from the James Jackson Bridge in Northwest Atlanta. Yellow-green fibres were found in his hair. Two days earlier Wayne Williams had been seen driving across this bridge, and he became a possible suspect in the murder case.

Williams' bedroom was found to contain a yellow-green carpet and thus a match was sought between fibres from this and those found on the murder victims.

It was determined that the carpet fibres were manufactured by the Wellman Corporation, and it was confirmed that no other fibre manufacturer was known to have made fibres similar in all respects – composition, cross-section and dye-type – to those found in Williams' carpet. Fibres of this type were sold only between 1967 and 1974. Prior to 1967, the company manufactured fibres of a similar composition but they had a different cross-section. Through numerous contacts with yarn spinners and carpet manufacturers, it was determined that the West Point Peperell Corporation had manufactured a line of carpet called Luxaire which was constructed in the same manner as Williams' carpet. One of the colours offered in the Luxaire line was called English Olive, visually the same as that of the Williams' carpet. It was now important to prove that the fibres on the victims were identical with those in Williams' carpet.

Several methods of investigation were used, following a logical sequence of tests. In all forensic science investigations, the non-destructive techniques should always be performed prior to any destructive analytical techniques. Only the minimum amount of sample that will give meaningful results should be used. This is important for two reasons. First, it allows a second analysis to be performed whenever possible. Second, it allows the 'other side' (defence or prosecution) to carry out their own investigation: this principle of equal opportunity is very important in forensic science. It also important that the sample taken is representative of the item in question.

The fibres for investigation having been selected, they were first examined by microscopy to determine their morphology, then subjected to UV/visible and vibrational spectroscopic analysis to reveal the type of fibre and the dyes used. Other techniques available are destructive to the sample and are used only if absolutely necessary. Thus thin-layer chromatography (TLC) and high-performance liquid chromatography (HPLC) are possible as further methods of dye identification, and **pyrolysis gas chromatography** (see Box 2.1) can also be used to characterize the polymer type.

BOX 2.1 Pyrolysis gas chromatography

Another common method for the analysis of fibres is **pyrolysis**. This is the high-temperature fragmentation of a molecule in an inert atmosphere. When a pyrolysis unit is coupled to a GC or GC–MS, a characteristic pattern, called a signature, or **pyrogram**, is obtained for different fibre classes. For example, acrylics break up into monomers, dimers, trimers and tetramers, with polyalkenes giving much the same fragmentation. The nylons and polyesters also give characteristic signatures. These signatures allow the generic class of the fibre to be identified, and the pyrograms of sample and control fibres to be compared.

Although two fibres may seem to have the same colour when viewed microscopically, the dyes used to achieve the colour could be compositionally distinct. Most textile fibres are impregnated with a mixture of dyes selected to give the desired shade. The significance of the fibre evidence is therefore enhanced when the forensic scientist can show that a questioned fibre (removed from a crime scene) and control fibre (from the source to be eliminated, such as the home of the suspect) actually have the same dye composition.

For known fibres, at least five spectra should be obtained for synthetic fibres and ten spectra for natural fibres, along the fibre length. This is to take into account the natural variation in the fibre due, for example, to differential uptake of the dyes, and also to take into account any damage which may have occurred to the fibre – for example twists, cracks or irregularities in the fibre shape.

The spectrophotometric analysis confirmed that the dyes used in the control fibres, removed from the carpet found in the home of Wayne Williams, and in the yellow-green fibres removed from the hair of Cater, were identical (Figure 2.2).

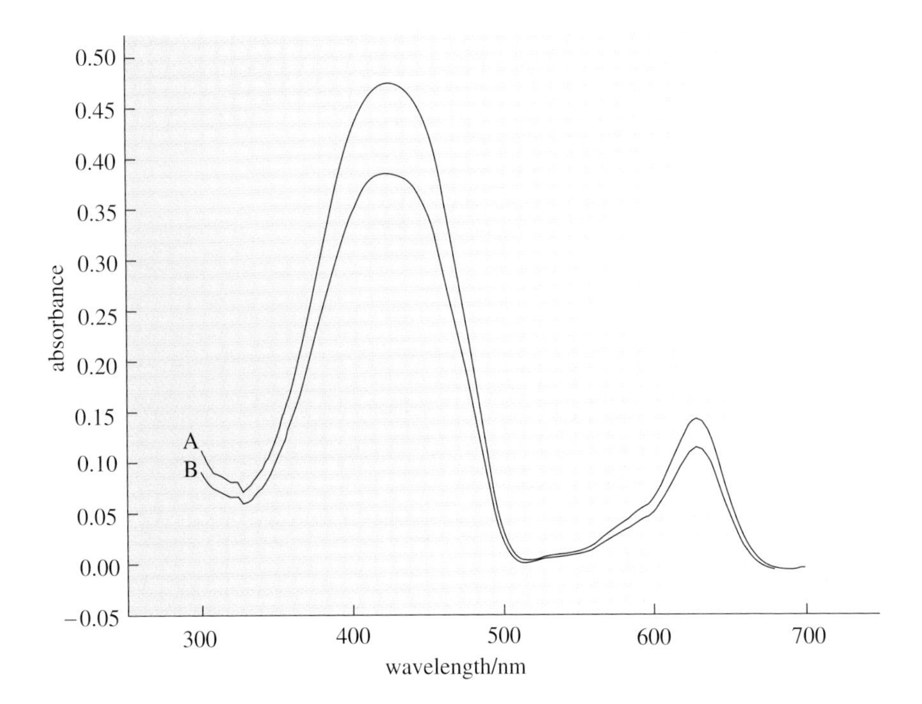

Figure 2.2
UV/visible spectra of the dyes in the fibres found A on the victims and B in Williams' carpet.

Infrared spectroscopy provides a powerful means of analysing fibres, and sample integrity can be preserved with the use of modern techniques. Most synthetic fibres are organic polymers, and like any organic compound their infrared spectra comprise absorptions from the characteristic functional groups present. Thus it is possible to identify major constituents, but more importantly for the forensic scientist the spectra from different types of fibre can be readily distinguished. Infrared spectroscopy clearly distinguishes between nylon (**2.1**) and acetate (**2.2**) fibres (Figure 2.3). Note the characteristic ester carbonyl stretch at 1 750 cm^{-1} for acetate, whereas the amide carbonyl stretch in nylon appears at a much lower frequency, around 1 690 cm^{-1}. The N—H stretch in nylon gives the strong peak at 3 300 cm^{-1}. The band at 3 500 cm^{-1} in the acetate spectrum is thought to be due to an overtone of the carbonyl stretch.

nylon-6,6, **2.1**

acetate, **2.2**

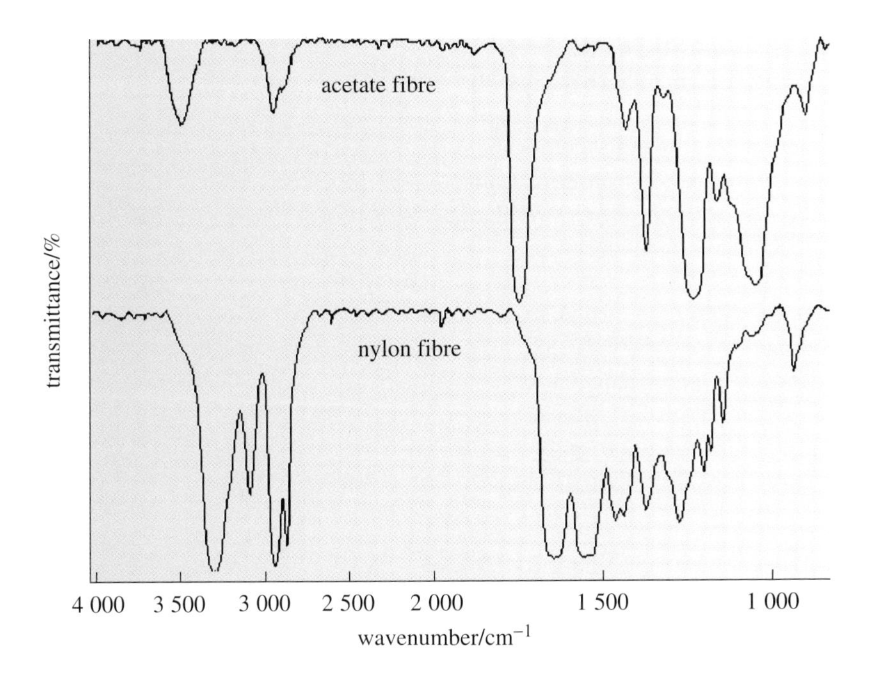

Figure 2.3
Infrared spectra of acetate fibres and nylon fibres.

For fibres of the same polymer from different manufacturers the 'fingerprint region' of the spectrum may often differ, if only subtly in some cases. Figure 2.4 shows a comparison of the spectra of the nylon fibres from Williams' carpet with those from a different manufacturer. As can be seen there are sufficient differences to allow a distinction to be made between them.

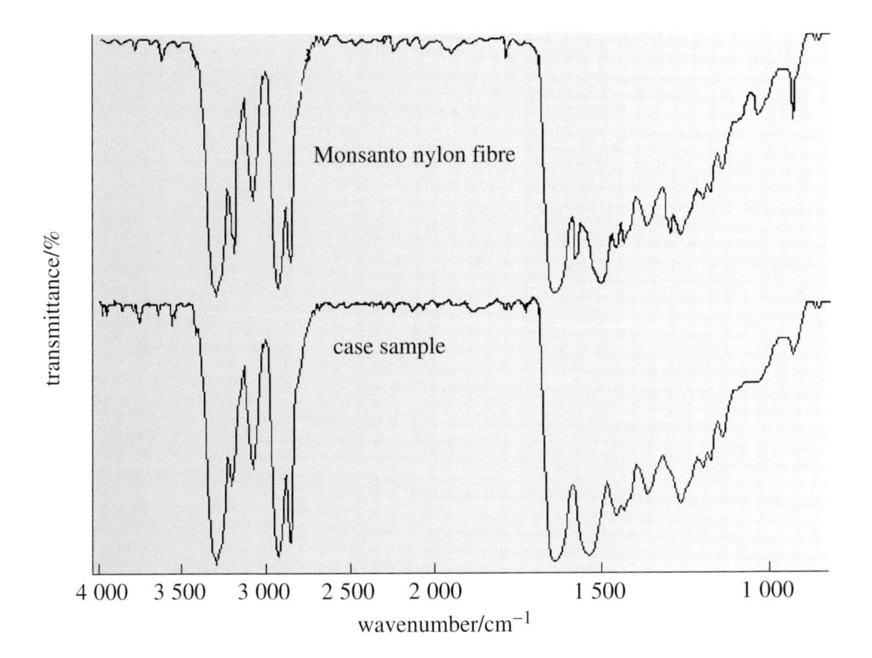

Figure 2.4
Comparative infrared spectra of nylon fibres.

The infrared results and those obtained from pyrolysis GC showed that the chemical compositions of the nylon fibres were the same. More specifically, all the fibres were identical with those produced by the West Point Pepperell Corporation and these were most probably Wellman 181B fibres. The microscopy of these fibres, and their unusual cross-section, all helped to show that they had originated from the same source, i.e. the carpet found in Williams' bedroom.

The value of fibres as evidence is related to the forensic scientist's ability to trace their origin. Often only a limited number of fibres need to be characterized, and the source identified. In summary, in the Williams' case the photoelectron micrograph indicated that the cross-section of the fibre was in fact rare, and would therefore aid the identification process. The spectroscopic analysis confirmed that the dyes used in the control fibres, removed from the carpet found in the home of Wayne Williams, and those used in the yellow-green fibres removed from the hair of Cater were chemically the same. The dye used was confirmed as English Olive, and in each case the composition of the dye was the same, as the UV/visible spectra were indistinguishable. The infrared results and those obtained from pyrolysis GC showed that that the chemical compositions of the fibres were the same. Close inspection of the spectra showed that the fibres had functional groups indicative of nylon fibres. More specifically, the fibres could not be discriminated from those fibres produced by the West Point Pepperell Corporation and were highly likely to be Wellman 181B fibres.

The West Point Pepperell Corporation had manufactured the Luxaire line from 1970 to 1975, but it had purchased the Wellman 181B fibre for this line only during the first year of production. This change of carpet fibre was yet another factor that made the Williams' carpet unusual. In order to quantify its unusual nature, statistical analysis was undertaken, something that had previously not been attempted in a case dependent on fibre-transfer evidence. It was estimated that a total of 16 397 square yards of carpet containing the Wellman 181B fibres dyed with English Olive was sold by the West Point Pepperell Corporation to retailers in ten States. By estimating the size of an 'average' room, and assuming that equal amounts of carpet were sold in the ten south-eastern States, it was estimated that 82 rooms with this carpet should be found in the State of Georgia. Using census details of the number of households and rooms, it was determined that the likelihood of finding another home in Atlanta with a room having carpet like Williams' was about 1 in 400 000.

This case helps to show a number of instrumental methods that can be widely applied in forensic science. This case was solved using only minute traces of fibrous material. Spectroscopy and chromatography demonstrate that fibres from the crime scene could be linked to Williams' home environment. Clearly, all the techniques were important as was the use of statistics to show that these particular fibres were very rare, thus adding strength to the case against Williams.

Williams has always maintained his innocence and is appealing against his conviction. He has been refused parole four times and his next parole review is due in 2005.

BOX 2.2 Trial report by John G Warner, UPI, 20 January 1982

The defense hammered away at the state's textile industry experts today, trying to get them to admit that the fibers found on the bodies of Wayne Williams' alleged victims could have come from anywhere.

'You do not pretend to tell the jury that you know where all this carpet ended up, do you,' defense lawyer Al Binder asked a representative of the firm which made the rug in Williams' home.

'Not at all,' responded Gene Baggett.

'As far as you know, it could have ended up in the Chattahoochee River, couldn't it?'

'As far as I know.'

The Chattahoochee River is where Nathaniel Cater, 27, and Jimmy Ray Payne, the two men Williams is on trial for murdering, were found.

Cater and Payne were among 28 young blacks abducted and murdered in Atlanta during a 22-month reign of terror that ended last spring when the 23-year-old would-be talent scout fell under suspicion. The state claims carpet and bedspread fibers found on their bodies match those from Williams' home.

Baggett, a representative of West Point Pepperell's carpet and rug division, was the only witness to testify this morning. Superior Court Judge Clarence Cooper declared a recess after his cross-examination.

The recess eventually turned into a lunch break when, according to court sources, the prosecution asked for a delay because its next witness was not ready.

Baggett also was the final witness Tuesday, testifying that a swatch of carpet – obviously removed from the Williams residence, although it has yet to be identified as anything but State's Exhibit 624 – appeared to be an example of that firm's Luxaire line.

That particular nylon plush line, he said, was sold for only a year – from December 1970 to December 1971. Only 1 555 yards of it were sold to private residences in the southeast.

The rug originated from tri-lobal fibers manufactured by Wellman Inc., witnesses testified.

Henry Poston, Wellman's director of technical services, said the fibers were, as far as he knew, unique. 'We know of no one else in the world who manufactured a fiber such as that,' he said.

Its singularity, he said, resulted from patent difficulties. Carpet makers wanted a fiber that had three lobes because it makes for more brilliant-appearing colors. But Du Pont had the patents sewn up on tri-lobals. So, Poston said, his firm devised a pattern which – unlike the identical lobes of the Du Pont patents – had two long lobes and one short one.

When the Du Pont patents expired in 1974, he said, Wellman promptly dropped the unique fiber and began manufacturing regular tri-lobal fibers.

Under cross examination, Poston admitted that he could not swear that no other manufacturer anywhere in the world made a similar lopsided fiber.

The bulk of the testimony came from a Du Pont expert, Herbert Pratt, a distinguished, gray-haired man whose boundless enthusiasm for textiles, their history and their manufacture led to long answers frequently illustrated on the blackboard.

It was Pratt who initially identified a fiber – presumably from one of the victims – as coming from the Wellman plant. But he admitted it was impossible to 'absolutely rule out' that the fiber did not come from some exotic overseas source.

CASE 2: POSSESSION OF HEROIN

3

3.1 The incident

During a police drugs raid, a quantity (about 100 g) of white powder was seized. Of the three people present in the room, two were dividing the powder at a table whilst the third was standing nearby. On questioning, the two seated at the table claimed that they were cleaning a spillage of caster sugar, and the third denied any knowledge of the substance. From information received, the police had reason to believe that heroin was being supplied (Box 3.1 and Figure 3.1a) and the white powder and the clothing of all involved were sent for forensic examination.

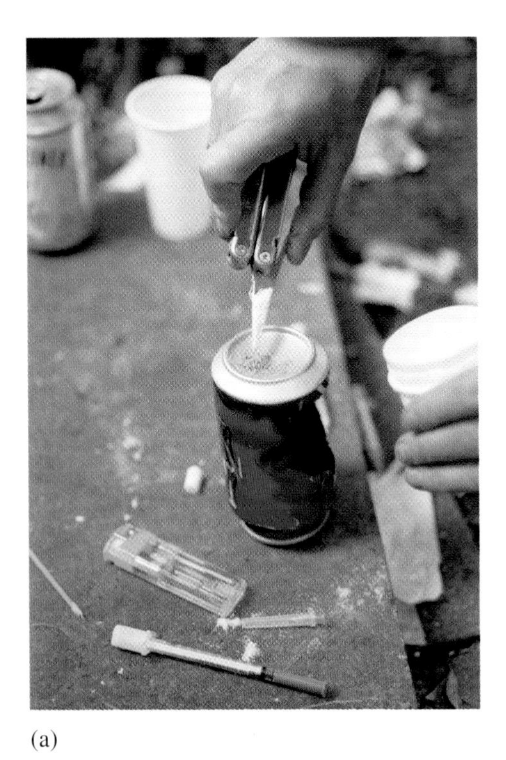

(a)

(b)

Figure 3.1
(a) Heroin powder;
(b) the opium poppy *Papaver somniferum*.

BOX 3.1 Opiate drugs

Opium is obtained as the milky resin from unripe seed capsules of the opium poppy (Figure 3.1b), which is grown mainly in India and Turkey. It was known as long ago as AD 100 as a folk medicine, but became a problem only in the 17th century when opium smoking became popular in China. Eventually addiction became such a problem that the Chinese tried to prevent all opium production and trade. Opium sets to a pale-brown, soft, sticky substance, which contains as its principal component the alkaloid **morphine** (**3.1**). This is the chief painkiller used in medicine, acting on the deep aching pain known as slow pain. **Codeine** is made from morphine by converting the —OH group on the benzene ring into the acetyl group, $-O-CO-CH_3$; this group is hydrolysed back to —OH in the body, producing morphine.

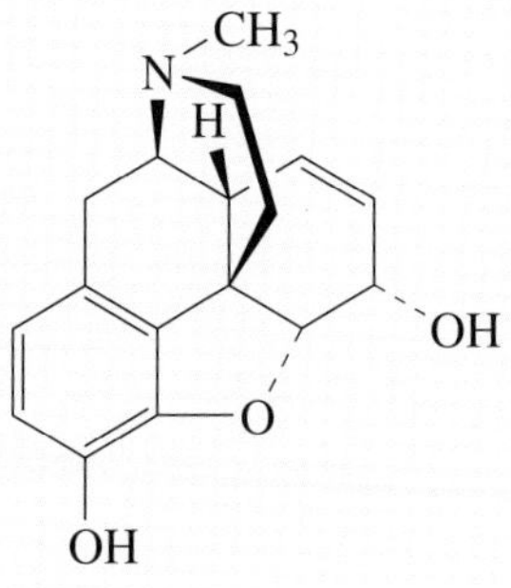

morphine, **3.1**

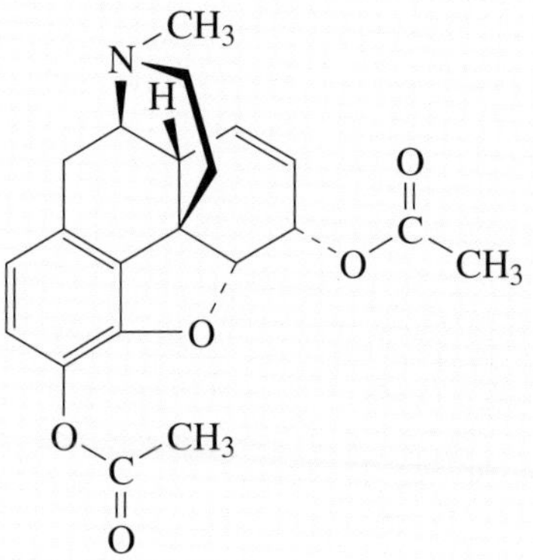

heroin, **3.2**

Heroin (**3.2**), a derivative of morphine sometimes known as diacetylmorphine or **diamorphine**, was first synthesized in 1898 and found to be several times more potent than morphine. Both —OH groups have now been converted into acetyl groups; this enables the heroin to be more soluble in the fats of the body, and so it passes more quickly through the blood–brain barrier. Once in the body the acetyl groups are again hydrolysed to —OH, producing morphine. Heroin was initially used as a cough suppressant until its highly addictive nature was recognized.

It is thought that opiates act in a similar way to the **enkephalins**, the painkillers produced naturally in the body, occupying the same nerve receptor sites and thus having the same analgesic effect. The opiates give a rush of pleasure and euphoria, but with continued use, the body demands larger and larger amounts. Withdrawal is extremely unpleasant, commonly accompanied by nausea, sweating, cramps, malnutrition, respiratory problems and low blood pressure. Recovery from addiction takes years of social and psychological rehabilitation.

3.2 The investigation

Chromatographic methods are widely used in forensic science because they are sensitive, quick and reliable. For example, illicit street drugs may be diluted with practically any material that is at the disposal of the drug dealer. The task of identifying the active component in an illicit drug preparation is made relatively easy by use of extraction, TLC, UV/visible, GC and GC–MS (Box 3.2). Knowledge of the purity of drugs can be useful for police intelligence purposes, because drugs that contain similar amounts of different diluents may be linked and provide information about how drugs are distributed. Similarly, the forensic toxicologist can use these chromatographic and spectroscopic methods to devise an analytical scheme that will successfully isolate, detect and specifically identify a toxic substance in body fluids.

BOX 3.2 Gas chromatography–mass spectrometry (GC–MS)

Gas chromatography is one of the most important analytical techniques because its ability to separate the components of a complex mixture is unsurpassed. However, GC does suffer one important drawback – its inability to produce a specific identification. A forensic chemist cannot unequivocally identify a substance based solely on a retention time determined from GC. If, however, the gas chromatograph is coupled to a mass spectrometer the problem can be overcome. The separation of the components of a mixture is firstly accomplished by GC. The direct connection to the mass spectrometer then allows a fragmentation pattern for each component to be obtained; as each mass spectrum is unique to each compound, this serves as a specific test for identifying that substance. It is also sensitive to minute concentrations.

At present, GC–MS finds its widest application in areas relating to the identification of drugs (including the routine testing of racehorses and greyhounds at racetracks). However, further research is expected to yield significant applications with respect to the identification of other types of physical evidence, including paint and greasy stains, such as shoe polishes and cosmetics.

In this drugs case, initial screening of the bulk sample was carried out by qualitative chemical tests commonly called **spot tests**. These tests are very useful for rapidly eliminating particular classes of drugs from consideration. The chemistry of the spot tests for barbiturates and for heroin is outlined in Figures 3.2 and 3.3, respectively. A barbiturate deprotonates on coordination to the cobalt ion in basic solution. The resulting complex has the characteristic intense violet-blue colour of many tetrahedral Co^{2+} complexes, such as $CoCl_4^{2-}$ or $CoCl_2(PR_3)_2$. In this case, testing with a methanolic mixture of Co^{2+} and 2-aminopropane did not give the characteristic colour, and thus an entire family of illicit drugs was excluded. The presence of morphine and its derivatives was tested for by adding a small amount of the powder to a dilute solution of methanal in sulfuric acid. The rapid development of the purple colour characteristic of heroin-based drugs gave a positive test in this case.

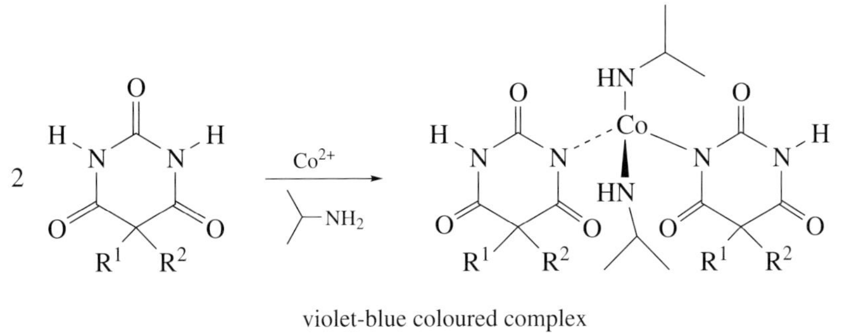

Figure 3.2 The production of the blue colour for barbiturates with Co^{2+} in a basic medium.

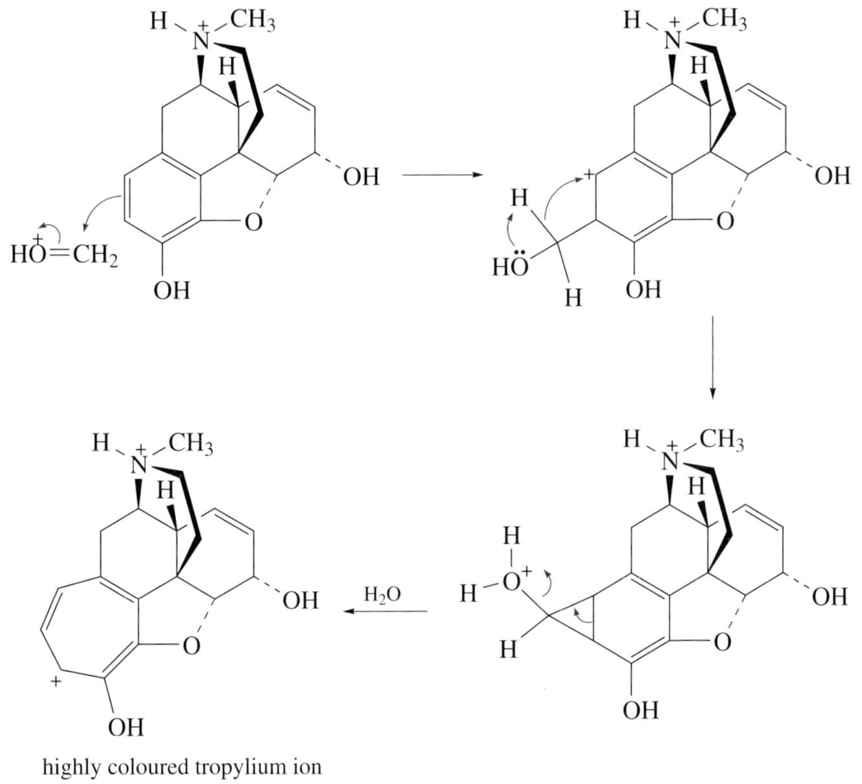

Figure 3.3 The production of the purple tropylium ion from acidified methanal and heroin.

Spectroscopic techniques were used to confirm the presence of heroin. The UV absorption spectrum showed a band at 280 nm, which is characteristic of the π–π^* transition in heroin. Owing to the broad nature of bands from electronic transitions, this was not sufficiently specific to allow unambiguous identification.

Analysis by infrared spectroscopy served only to confirm that the sample was a mixture, as the spectrum did not correspond to that of any single pure substance in the database. As organic compounds have many similar absorptions, it proved difficult to obtain a clear identification in this case.

With the large amount of sample confiscated, analysis by TLC was possible in this case. A small amount of the powder was dissolved in methanol, spotted onto a TLC plate and eluted with methanol. Standards were also spotted onto the same plate for comparison. The plate, shown in Figure 3.4, has the case sample, standard pure heroin, sugar, and quinine (a common diluent for heroin). After development of the plate, two spots were seen from the case sample, which by comparison with the retention times of the standards, corresponded to heroin and quinine. The bulk sample was thus proved to contain heroin mixed with quinine. No sugar was detected by any of the tests.

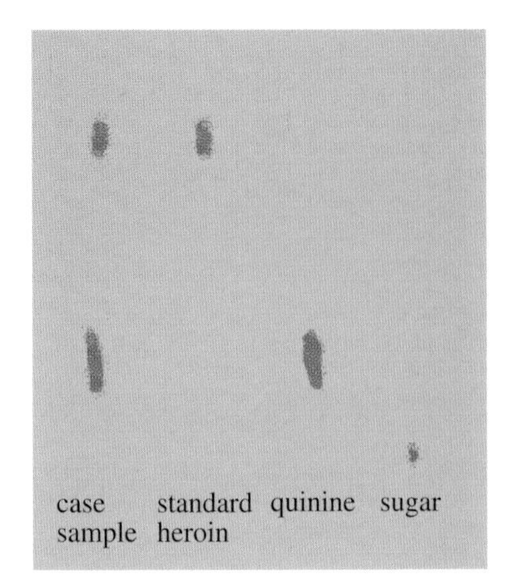

Figure 3.4
Thin-layer chromatography plate of the bulk sample.

Examination of the clothing of the man standing by the table revealed traces of white powder adhering to the fibres of the inside jacket pocket. The particles were removed and submitted for analysis, with the suspicion that they too were heroin.

Analysis of such small amounts of materials requires particularly sensitive instrumental techniques, which should ideally be able to provide separation from any contaminants and an unambiguous identification. GC–MS is the ideal tool for this as this not only provides a separation from any impurities, but also gives an absolute identification in the mass spectrum.

The sample solution was subjected to GC–MS, which automatically records the mass spectrum of each compound as it emerges from the column. Modern instruments have a searchable database of mass spectra, which allows the best match between the case sample and known substances to be found very quickly. The results in this case are shown in Figure 3.5. Here the best match with the mass spectrum identified by the computer is with heroin. Visual examination by the human operator (still an essential part of the process!) confirms that the two spectra are virtually identical. There are small differences between the two spectra, but these are accounted for by noise in the sample spectrum (probably due to the very low concentration used) and the fact that the library spectra were probably recorded on a different instrument. The other heroin derivatives found from the database, have obvious differences in their mass spectra and can be ruled out.

All three suspects were eventually convicted of dealing in dangerous drugs.

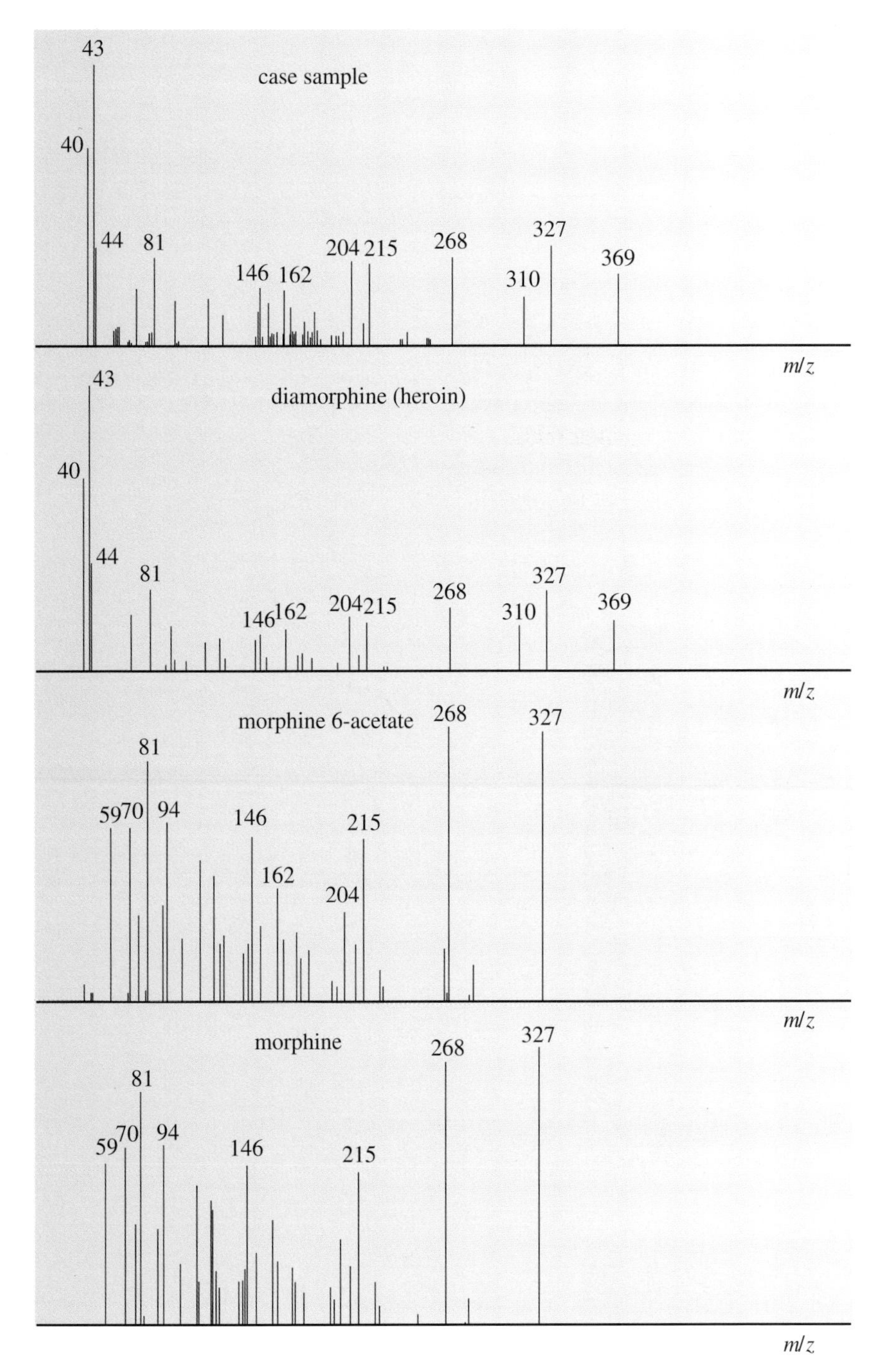

Figure 3.5 Comparison of the mass spectra obtained from the case sample and library standards.

CASE 3: THE ATTEMPTED MURDER OF ALAN SMITH *

4

4.1 The incident

Mr Smith, a bank manager, was driving to work at about 7.30 a.m. on the morning of Monday 14 October 1996, when his car swerved off the road and crashed into a ditch. Mr Smith was badly injured in the crash and did not fully recover for a number of months. Several eye witnesses reported hearing a 'loud bang' and seeing dense white smoke coming from the car. The area was sealed, and police forensic scientists examined the wreckage of the car and the damage to the road surface in the vicinity of where the explosion was thought to have occurred (Figure 4.1).

Figure 4.1 The aftermath of a car-bomb.

Figure 4.2 A battle re-enactment involving the use of blackpowder.

* Most bombing incidents are politically sensitive, so this case is fictional.

4.2 The investigation

After research into the business background of Mr Smith's local branch of the bank, a list of possible suspects who might have had a motive for the attempted murder was drawn up. After preliminary enquiries Mr Jones, the owner of a toy and model shop, was thought to have both a motive (the bank had recently filed for the bankruptcy of Mr Jones' shop and was in the process of repossessing his home) and the opportunity, having no supportable alibi for his actions the previous weekend. A search of Mr Jones' premises revealed quantities of 'blackpowder', a commercial explosive. However, Mr Jones stated that one of his hobbies was participating in re-enactments of historic battles (Figure 4.2) and that this hobby required the use of considerable quantities of blackpowder.

Blackpowder is the name given to a commercially available form of gunpowder. Its composition by weight is 75% potassium nitrate (KNO_3), 15% charcoal (C), and 10% sulfur, (S). Explosions can be put into two broad categories: 'deflagration' or 'low-order', where the decomposition occurs at a speed equal to or less than the speed of sound, usually less than 2 000 m s^{-1}; and 'detonation' or 'high-order', where the speed of decomposition is greater than the speed of sound within the compound and can reach velocities up to 8 500 m s^{-1} in compounds such as nitroglycerine, TNT, and RDX. Blackpowder is a low-order explosive. Both high and low explosives are dangerous when ignited in confined spaces as both rapidly release large amounts of gases. In any explosion it is very rare for all the charge to be consumed. Generally traces of uncombusted material are scattered by the blast and deposited on nearby objects. The recovery and analysis of uncombusted explosive is the role of the forensic explosives investigator, who may use a wide variety of techniques to identify the explosive used.

For blackpowder, the equation representing its explosive decomposition is:

$$3C(s) + S(s) + 2KNO_3(s) \longrightarrow 3CO_2(g) + N_2(g) + K_2S(s) \qquad (1)$$

Initial screening of the area was carried out with an EGIS detector. This samples the air for volatile organic compounds, traces of which remain from organic-based high explosives, and has to be carried out as soon after the explosion as possible. The vapours are screened by GC followed by pyrolysis of each component as it leaves the first GC column. The pyrolysis products are then passed through a second GC column, and the peak profile of retention times and relative intensities gives a characteristic signature for a particular residue. Further tests in the laboratory on known standards can give proof of the presence of an individual explosive. In this case the EGIS detector found no traces of organic vapours from explosives, only residual hydrocarbons from petrol and oil.

Evidence recovered from the scene included powder residues from the car itself and a crater in the road. These were thought to have originated from the explosive and were subjected to scientific investigation.

The infrared spectra shown in Figure 4.3 were obtained from collections of small particles of powder and are representative across about 50 such samples recovered from the scene; they are compared with the infrared spectra of a sample of the local soil and potassium nitrate. Several features were assigned to the presence of substances that would be found under normal circumstances: for instance, the strong band at 1 100 cm^{-1} was assigned to the Si—O stretch in silicates from soil, and the features above 3 000 cm^{-1} were thought to be due to the O—H stretch from moisture adsorbed to the surface of the soil.

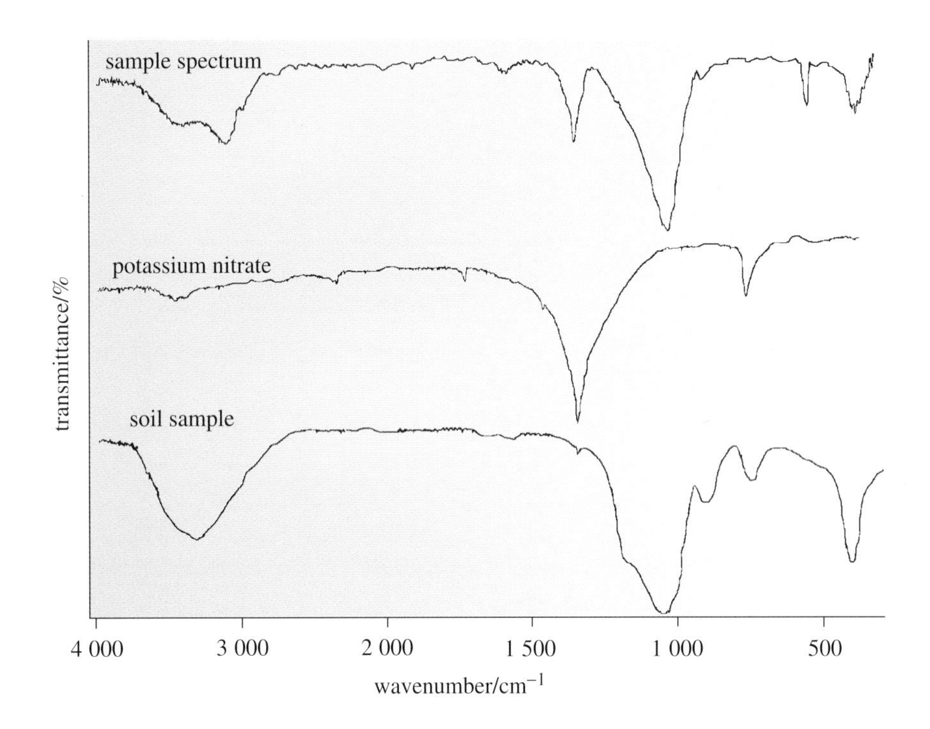

Figure 4.3
Comparative infrared spectra.

The band at about 1 400 cm^{-1} was assigned to nitrate. The lower than expected frequency of the O—H stretch could reasonably be attributed to different extents of hydrogen bonding of the adsorbed water to the silicates in the soil.

Initial conclusions were that the match provided reasonable evidence that 'blackpowder' was involved in the explosion. A link between Mr Jones and the crime had thus been established and he was charged.

But was this really conclusive? Do the spectra match sufficiently closely? What about the peaks at about 800 cm^{-1} in KNO_3 and soil? Why aren't they present in the spectrum from the case sample? What about the peak at about 600 cm^{-1} in the case sample, how can this be explained?

Fortunately for Mr Jones, senior scientists felt that the analysis of the residues was not complete, and that the differences in the frequencies between the spectra obtained from the crime scene and standards, whilst small, were still, perhaps, significant. They therefore carried out other tests and comparisons. In particular, they ran Raman spectra of the recovered materials and also of reference samples in an effort to detect traces of sulfur and charcoal which should be present in residues from blackpowder. These spectra were obtained using a 'Raman microscope', which has sufficient resolution to obtain spectra from individual particles within the sample. The particles recovered from the explosion scene gave spectra summarized in Table 4.1.

Comparison of the spectra obtained with reference spectra indicated that neither the nitrate ion nor sulfur was present, as some characteristic features are absent in the spectra from the residues. The results suggested that blackpowder was not the explosive used.

These results led the forensic scientists to propose that a different type of explosive had been used, which was not nitrate-based. Ammonium perchlorate (NH_4ClO_4) is a strong oxidant, and when mixed with aluminium powder, or other reducing agents, forms a particularly potent explosive (a mixture of aluminium and ammonium perchlorate is used as the solid fuel for the US space shuttle). The Raman spectra of KNO_3 and NH_4ClO_4 (Figure 4.4) clearly show the differences between the two compounds.

Table 4.1 Raman spectra of particles (A and B) recovered from the crime scene, and of reference substances

Substance	Wavenumber/cm^{-1}						
particle A			734				
particle B		623		936(s)[a]	1067	1111	
potassium nitrate, KNO_3 (NO_3^-)			716(ir)[b]		1049(s)		1390(ir)
ammonium perchlorate, NH_4ClO_4 (ClO_4^-)	459	625(ir)		928(s)		1119(ir)	
ammonium perchlorate, NH_4ClO_4 (NH_4^+)							1400(ir) 1680 3040 3145(ir)
sulfur, S_8			815 868				

[a] Most intense peak marked (s) for strong [b] Mode active in the infrared.

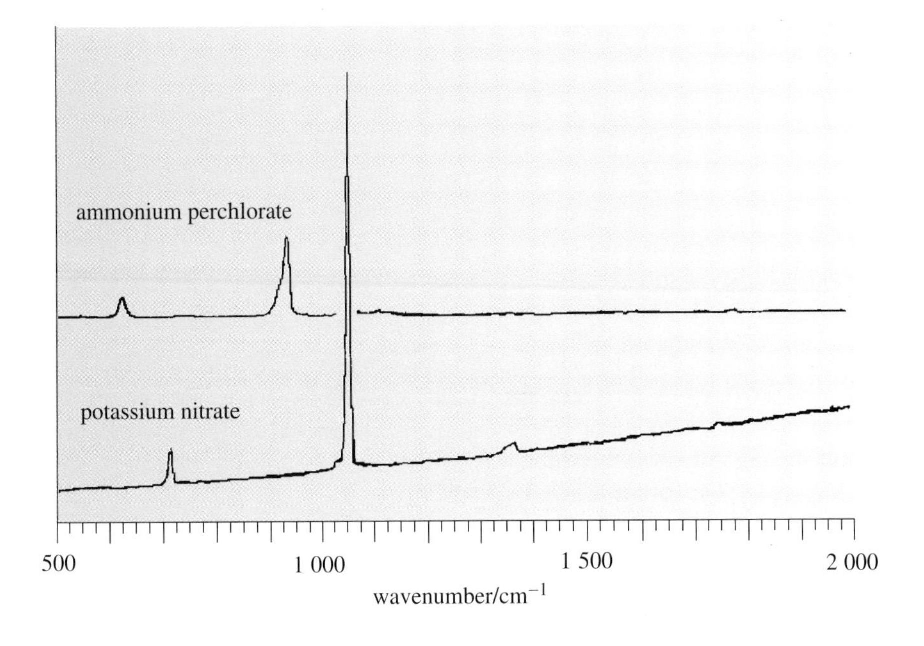

Figure 4.4
Comparative Raman spectra.

The Raman spectra of the particles were assigned by comparison with authentic samples:

- Type A: silica particles from the soil.
- Type B: residual ammonium perchlorate, which has an intense band at 934 cm^{-1}.

To give further proof, these conclusions must also explain the observed infrared spectra. The infrared spectrum of ammonium perchlorate has bands at 3 145 cm^{-1} due to N—H stretching, 1 400 cm^{-1} due to bending modes in the NH_4^+ ion, and 1 120 cm^{-1} due to Cl—O stretching modes. Thus the original initial incorrect assignment of nitrate was due to the proximity of the N—O stretch in nitrate (1 390 cm^{-1}) and the N—H bending mode in ammonium (1 400 cm^{-1}) with the Cl—O stretching being masked by the Si—O stretch in the silicates of the soil.

The use of infrared spectroscopy alone gives ambiguous results in this case: a combination of infrared and Raman spectroscopy was critical in establishing the identity of the recovered evidence.

Once the new interpretation was presented it was decided not to proceed with charges against Mr Jones, and the investigation was continued for a new suspect.

CASE 4: THE *HINDENBURG* DISASTER

5

5.1 The incident

The airship *Hindenburg* was due to arrive at the Lakehurst Naval Air Station, New Jersey, at 6 a.m. on 6 May 1937. It was delayed by about 12 hours by strong headwinds over the Atlantic and finally approached from the north in the late afternoon just as a thunderstorm was gathering in the west. It detoured towards Atlantic City and started to make its final approach to the mooring tower from the south-west at 7.00 p.m. in light rain, then disaster struck (Figure 5.1): 36 of the 92 passengers and crew were killed.

BOX 5.1 Airships

Airships or dirigibles consist of a bag filled with a gas that is lighter than air, provided with a means of propulsion and steering, and a gondola for passengers and crew. The gas used is either hydrogen or helium. The first successful airship, designed by Henri Giffard, flew in 1852. There were three main types of airship, with non-rigid (blimps), semi-rigid or rigid gas bags.

The first regular commercial passenger travel was provided by the German Zeppelin airships, starting in 1910. Many countries had developed their own ships for both commercial and military use, but the loss rate of these ships was quite severe. Britain lost both the *R34* and the *R38* in 1921, and the USA lost the *Roma* in 1922 and the *Shenandoah* in 1925. In 1929 Britain resumed construction of the rigid ships, launching the *R100* and the *R101*. In 1930 the *R101* crashed in France in a violent storm whilst on the way to India, killing 46 people. After this the British government decommissioned the *R100* and abandoned any further construction of dirigibles. Interestingly, none of the above incidents was directly caused by the use of hydrogen as lifting gas, although leaking hydrogen ignited in the *R38* and *R101* crashes and undoubtedly contributed to the loss of life; American airships used helium for buoyancy.

The *Hindenburg*, the pride of the Third Reich, at over 800 feet in length, is still the largest object ever to have flown. Together with its sister ship, the *Graf Zeppelin*, it had completed over one million miles in passenger service without incident, and the *Hindenburg* itself had made 21 Atlantic crossings. With a cruising speed of about 80 m.p.h. it provided a luxurious and rapid passage across the Atlantic. It is a testament to German engineering skills that they had avoided the tragedies that plagued the British and American airship programmes.

Since the destruction of the *Hindenburg*, airships have been of only the non-rigid variety and are mainly used for military and advertising purposes.

Figure 5.1
(a) A diagram of the internal structure of the *Hindenburg*.
(b) The *Hindenburg* intact; it was 825 feet long.
(c) The airship bursts into flames.

The demise of the *Hindenburg* has attracted particular interest for a number of reasons. The arrival of the *Hindenburg* was newsworthy in its own right, and its spectacular end was not only the subject of a radio broadcast (see Box 5.2), but was also captured on film.

BOX 5.2 Radio broadcast by Herbert Morrison

Herbert Morrison, a correspondent with Chicago's WLS radio station, along with his sound engineer, Charlie Nehlson, witnessed the terrifying demise of the *Hindenburg*. There followed one of the most famous broadcasts in history: a transcript follows.

'Well, here it comes, ladies and gentlemen; we're out now, outside of the hangar. And what a great sight it is, a thrilling one, just a marvellous sight. It's coming down out of the sky, pointed directly towards us and toward the mooring mast. The mighty diesel motors just roared, the propellers biting into the air and throwing it back into a gale-like whirlpool. No wonder this great floating palace can travel through the air at such a speed, with these powerful motors behind it.'

'Now and then the propellers are caught in the rays of sun, their highly polished surfaces reflect....The sun is striking the windows of the observation deck on the eastward side and sparkling like glittering jewels on the background of black velvet.'

'Now the field, as we thought active when we first arrived, has turned into a moving mass of cooperative action. The landing crews...their posts...and orders are being passed along, and last-minute preparations are being completed for the moment we have waited for so long.'

'The ship is riding majestically toward us like some great feather, riding as though it was mighty good...mighty proud of the place it's playing in the world's aviation. The ship is no doubt bustling with activity as we can see; orders are shouted to the crew, the passengers probably lining the windows looking down at the field ahead of them, getting their glimpse of the mooring mast. And these giant flagships standing here, the American Airline flagships, waiting to direct them to all points in the United States when they get the ship moored.'

'There are a number of important persons on board, and no doubt the new commander, Captain Max Pruss, is thrilled, too, for this is his great moment, the first time he's commanded the *Hindenburg*. On previous flights, he acted as Chief Officer under Captain Lehmann.'

'It's practically standing still now. They've dropped ropes out of the nose of the ship, and it's been taken a hold of down on the field by a number of men. It's starting to rain again; the rain had slacked up a little bit. The back motors of the ship are just holding it, just enough to keep it from — .'

'It burst into flames! Get out of the way! Get out of the way! Get this, Charlie! Get this, Charlie! It's fire and it's crashing! It's crashing terrible! Oh, my! Get out of the way, please! It's burning, bursting into flames and is falling on the mooring mast, and all the folks agree that this is terrible. This is the worst of the worst catastrophes in the world! Oh, it's crashing...oh, four or five hundred feet into the sky, and it's a terrific crash, ladies and gentlemen. There's smoke, and there's flames, now, and the frame is crashing to the ground, not quite to the mooring mast...Oh, the humanity, and all the passengers screaming around here!'

'I told you...I can't even talk to people...around there. It's — I can't talk, ladies and gentlemen. Honest, it's just laying there, a mass of smoking wreckage, and everybody can hardly breathe and talk...I, I'm sorry. Honest, I can hardly breathe. I'm going to step inside where I cannot see it. Charlie, that's terrible. I — Listen folks, I'm going to have to stop for a minute, because I've lost my voice...This is the worst thing I've ever witnessed......'

5.2 Original investigation

Following the disaster several lines of investigation were pursued.

Owing to the highly charged political climate of the time, speculation about sabotage against the Third Reich was rife and the FBI conducted a lengthy investigation into this possibility. The documents from that investigation are now released under the US Freedom of Information Act (Figure 5.2).

Figure 5.2 Document from the original FBI investigation.

JOHN EDGAR HOOVER
DIRECTOR

Federal Bureau of Investigation
United States Department of Justice
Washington, D.C.

PEF:RP

May 10, 1937

ALL INFORMATION CONTAINED
HEREIN IS UNCLASSIFIED
DATE 9-15-__ BY SPIAC__

MEMORANDUM FOR THE DIRECTOR

RE: HINDENBURG DISASTER

Mr. Connelley called with reference to this matter and stated that Commander Rosendahl has been very favorable to us as has the Department of Commerce Committee. He stated that there are two representatives of Senator Copeland's committee at Lakehurst, Colonel Hartney and Roger Williamson, Hartney having been appointed as Technical Adviser. Mr. Connelley stated that he does not believe Commander Rosendahl or the Commerce Committee is particularly favorable to them, but they have not taken any action against them except to listen to their suggestions. Mr. Connelley said that the first thing they had in mind was the foot tracks which came in from the back gate on the west side of the reservation. There is a road running along there to which the public has access but there is a barbed wire fence between the road and the reservation itself. At the time of the landing of the ship there were many spectators and automobiles outside of the fence, but after the crash the general public swarmed over the field from all directions and it was some three or four hours before a military patrol was established and the public was excluded. These tracks were found leading from this gate in the back where two persons had undoubtedly climbed over the fence and walked into the reservation. Mr. Connelley stated that it was first believed that these tracks were made by two boys who were picked up on the seventh, but it was further learned that the boys have no part in the picture and that they did not leave the tracks in question. The tracks were apparently made by some of the people who swarmed over the field. However, Williamson has talked with Senator Copeland and has put emphasis on the fact that these prints are there and evidently Senator Copeland has become interested in it. Mr. Connelley stated that they have photographed the prints and also taken plaster casts of them and that we could undoubtedly make an identification if we could find the persons to whom they belonged. However, Mr. Connelley stated that there must be numerous other tracks of this kind in every direction from the reservation, and he does not believe that they bear any significance but again they might lead to someone who went to the post and did something there. Mr. Connelley stated that one of the ideas advanced is that somebody went to the post and possibly shot the ship down as it has been indicated that a survey is being made of the ground with a possibility of finding some empty shells.

62-48190-2

Mr. Connelley stated that Commander Rosendahl has been approached with the suggestion that he assign a number of enlisted personnel to make a very careful check of the field and it was suggested that we be put in charge of the investigation

COPIES DESTROYED
10 OCT 16 1964

5

No definite proof of sabotage was ever found, although numerous threats against the airship had been made.

A combination of structural damage and lightning strike was also proposed. The *Hindenburg* underwent a series of sharp turns at full speed before the attempted mooring, and it was thought that bracing wires may have broken under the stress puncturing a gas cell, the ignition of the escaping hydrogen then causing the disaster. (A similar fate befell the *R38* which broke in two during a sharp turn with the front section subsequently exploding.) Although no conclusive evidence was ever obtained, the explanation that lightning had ignited a hydrogen leak was widely held to be the cause of the *Hindenburg* disaster.

5.3 Recent scientific investigation

More recently, a retired NASA scientist, Addison Bain, proposed that several features of the incident were inconsistent with a hydrogen leak.

- The photographs of the early stages of the fire show that the airship was still level. If the fire was due to burning hydrogen, there would be a loss of buoyancy in the tail section leading to a 'nose-up' attitude. This is very apparent in the later stages of the fire when the tail gas cells had definitely ruptured.
- The *Hindenburg* did not explode, but burned rapidly in all directions.
- Falling pieces of fabric were burning and not self-extinguishing.
- Although all pictures of the incident were taken in black and white, eyewitness accounts have allowed a 'colouring' of the original photographs to be done. These reports all state that the flames were an intense orange-yellow in colour. This is inconsistent with a pure hydrogen flame, which is almost colourless. (However, it could well be due to the presence of salt on the outside covering, as the *Hindenburg* had just flown across the Atlantic.)
- No major gas leak was detected on board the *Hindenburg* itself, where the pressure sensors on the gas cells had all registered their normal readings throughout the flight.
- No-one smelled garlic, even though this scent had been added to the hydrogen to help to detect a leak.
- The airship made a 'high' landing, and was winched down using landing lines – this would make a ground-to-cloud electrical path in the highly charged atmosphere.

The proposed explanation of the fire focused on the nature of the covering of the craft. Although exact details of the original construction are not known, several fragments of the fabric from the *Hindenburg* survived the fire, and were available for analysis. The infrared spectra showed that inorganic nitrate and cellulose acetate (Figure 5.3) were used as conditioners for the fabric for strengthening and water-proofing. Also, to avoid excessive heating of the gas cells, the outer skin was also covered with a layer of highly reflective aluminium metal and iron oxide. These two components are not detectable by infrared spectroscopy but can be analysed directly by scanning electron microscopy.

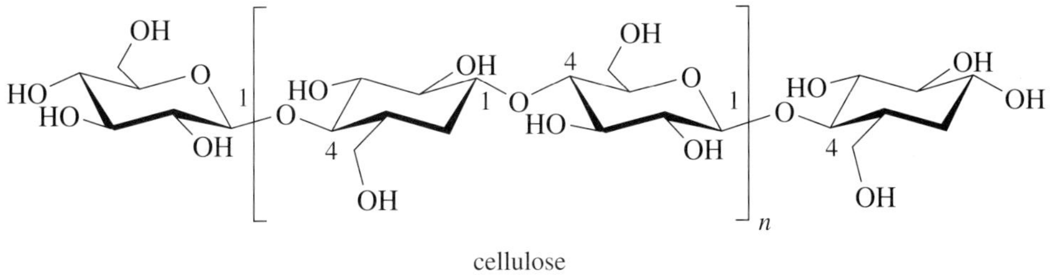

cellulose

Figure 5.3 The structure of cellulose. Cellulose acetate is a polymer made by substituting one, two or three (but most commonly two) of the $-OH$ groups on the ring with an acetyl, $-OCOCH_3$, group.

The importance of the reflective coating can be understood by simple calculations. The *Hindenburg* had a hydrogen gas capacity of about 200 000 m^3. Assuming this behaves as an ideal gas, the volume change if the temperature of the gas were to rise by 5 °C from 298 K to 303 K is given by

$$\frac{200\,000}{298} = \frac{V_2}{303} = 203\,356\,\text{m}^3$$

an increase of more than 3 000 m^3!

The mixture of materials used is a highly combustible cocktail of oxidants and fuel. The components of the reflective paint, iron oxide and aluminium, under the right circumstances will react together violently:

$$2Al(s) + Fe_2O_3(s) = 2Fe(s) + Al_2O_3(s) \quad (2)$$

The combination of cellulose acetate and nitrate ion has similar properties, with nitrate being a good oxidant and cellulose-based materials a good fuel (wood is largely cellulose-based). Normally, there would be no reason for any of these components to ignite spontaneously. Even in a lightning strike the charge would be safely spread over the entire surface of the ship, being harmlessly earthed, for example by the landing ropes on mooring. However, an eyewitness had reported seeing a blue glow of electrical activity dancing on top of the *Hindenburg* before the fire started. This led Bain to propose that an electric charge built up on the craft as it flew in the vicinity of thunderstorms in the area. The grounding of this charge by the mooring ropes would cause sparking between an earthed and an adjacent unearthed panel, which could set the covering fabric on fire. Normally, all the panels would be electrically connected, but the potential difference between a single unearthed panel and the grounded adjacent sections would be sufficient to cause sparking across the gap. The resulting fire would then spread rapidly along the ship, bursting the gas cells and igniting the hydrogen in the process.

To test this theory a series of experiments was carried out on samples of fabric of similar composition to those believed to have been used on the *Hindenburg* and finally on samples of the original fabric itself, and it was found that:

- the behaviour of the fabric under simulated lightning strike conditions (with the discharge perpendicular to the surface) failed to cause ignition;
- an electrical discharge along the surface of the fabric had a very different result, initiating a vigorous reaction, and rapidly destroying the fabric.

The colour of the flame does not completely prove that hydrogen was not involved in the initial fire. The *Hindenburg* had flown over the Atlantic at low altitude (by today's standards) at about a thousand feet, and would certainly have had salt coating its outer skin. Similarly, the inorganic nitrate used to dope the outer skin would probably contain sodium or potassium ions. Almost any sort of flame is sufficiently hot to produce the characteristic, very strong emission spectrum from sodium atoms, giving the vivid orange-yellow colour. Thus although there is no reason to doubt the eyewitness accounts of the colour of the flame, it is consistent with either a hydrogen fire burning through the skin of the craft or a fire of the skin alone.

All the evidence thus seemed to point Bain towards an *aluminium* fire in the fabric of the *Hindenburg*.

A final piece of corroboration came when a letter was uncovered in the Zeppelin Archive in Friedrichshafen, Germany, handwritten by an electrical engineer, Otto Beyersdorff. On 28 June 1937, he wrote:

> The actual cause of the fire was the extreme easy flammability of the covering material brought about by discharges of an electrostatic nature.

ACKNOWLEDGEMENTS

Grateful acknowledgement is made to the following sources for permission to reproduce material in this book:

Figure 3.1a: Photography Library; *Figure 3.1b*: Oxford Scientific Films; *Figure 4.1*: Associated Press; *Figure 4.2*: Courtesy of The Sealed Knot; *Figure 5.1*: Topham Picturepoint.

Every effort has been made to trace all the copyright owners, but if any has been inadvertently overlooked, the publishers will be pleased to make the necessary arrangements at the first opportunity.

INDEX

Note: Principal references are given in bold type; picture references are shown in italics.

A

B

C

D

E

CD-ROM INFORMATION

Computer specification

The CD-ROMs are designed for use on a PC running Windows 95, 98, ME, 2000 or XP. We recommend the following as the minimum hardware specification:

Processor	Pentium 400MHz or compatible
Memory (RAM)	32MB
Hard disk free space	100MB
Video resolution	800 × 600 pixels at High Colour (16 bit)
CD-ROM speed	8 × CD-ROM
Sound card and speakers	Windows compatible

Computers with higher specification components will provide a smoother presentation of the multimedia materials.

Installing the CD-ROMs

Software must be installed onto your computer before you can access the applications. Please run INSTALL.EXE from either of the CD-ROMs.

This program may direct you to install other, third party, software applications. You will find the installation programs for these applications in the INSTALL folder on the CD-ROMs. To access all the software on the CD-ROMs you must install Adobe Acrobat™ Reader.

Running the applications on the CD-ROM

You can access the *Separation, Purification and Identification* CD-ROM applications through a CD-ROM Guide (Figure C.1) which is created as part of the installation process. You may open this from the **Start** menu, by selecting **Programs** followed by **The Molecular World**. The CD-ROM Guide has the same title as this book.

The *Data Book* is accessed directly from the **Start | Programs | The Molecular World** menu (Figure C.2) and is supplied as an Adobe Acrobat document.

Problem solving

The contents of this CD-ROM have been through many quality control checks at the Open University and we do not anticipate that you will encounter difficulties in installing and running the software. However a website will be maintained at

http://the-molecular-world.open.ac.uk

that details solutions to any faults that are reported to us.

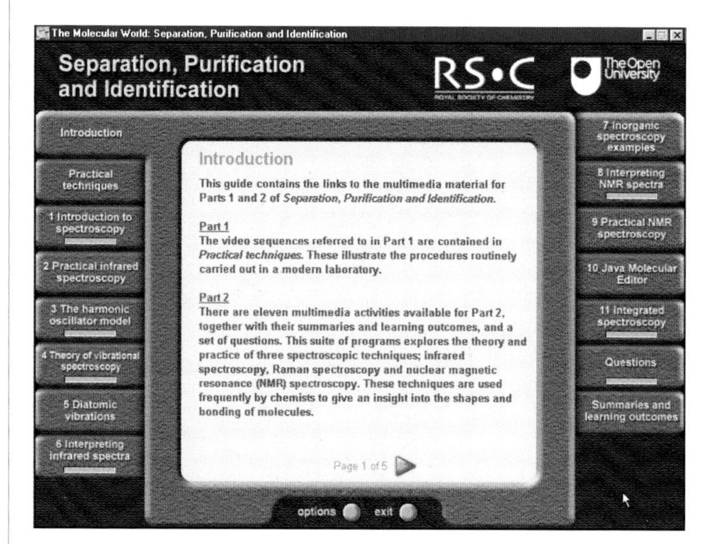

Figure C.1 Figure C.1 The CD-ROM Guide

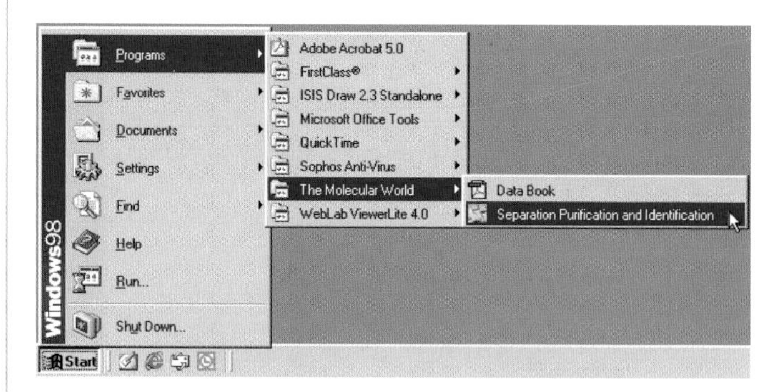

Figure C.2 Accessing the Data Book and CD-ROM Guide